杨力 著

Spark
大数据实时计算
基于 Scala 开发实战

人民邮电出版社

北京

图书在版编目（CIP）数据

Spark大数据实时计算：基于Scala开发实战 / 杨力著. -- 北京：人民邮电出版社，2022.10
ISBN 978-7-115-59703-8

Ⅰ．①S… Ⅱ．①杨… Ⅲ．①数据处理软件 Ⅳ．①TP274

中国版本图书馆CIP数据核字(2022)第121048号

◆ 著　　杨　力
　责任编辑　赵　轩
　责任印制　陈　犇

◆ 人民邮电出版社出版发行　北京市丰台区成寿寺路11号
邮编　100164　电子邮件　315@ptpress.com.cn
网址　https://www.ptpress.com.cn
北京七彩京通数码快印有限公司印刷

◆ 开本：787×1092　1/16
印张：19.5　　2022年10月第1版
字数：413千字　2025年1月北京第5次印刷

定价：79.80元

读者服务热线：(010)81055410　印装质量热线：(010)81055316
反盗版热线：(010)81055315
广告经营许可证：京东市监广登字 20170147 号

前言

今天，大数据和人工智能正以前所未有的广度和深度影响着各行各业。现在及未来公司的核心壁垒就是数据，核心竞争力来自基于大数据的人工智能。Spark 是当今大数据领域最活跃、最热门、最高效的大数据通用计算平台之一。

在任何规模的数据计算中，Spark 在性能和扩展性上都具有优势。

Spark 中的 Spark SQL、Spark Streaming、MLlib、GraphX、R 五大子框架和库之间可以无缝地共享数据和操作，这不仅打造了 Spark 在当今大数据计算领域其他计算框架都无可匹敌的优势，而且使 Spark 正在加速成为大数据处理中心首选的通用计算平台。

本书将带您了解 Spark 大数据实时计算的基本概念并进行实战操作。通过对本书的学习，您将对 Spark 大数据实时计算技术有深刻的认识，并且掌握大数据技术中主流的实时计算工具 SparkRDD、Spark SQL、Spark Streaming 等；再通过对大数据的实时计算项目案例开发的学习，您将了解 Spark 大数据实时计算技术的实际应用。学习本书是您掌握大数据实时计算技术非常好的入门途径。

作者在编写本书时力求内容科学准确、系统完整、通俗易懂，让初学者能快速掌握大数据技术，同时对专家级读者也具有一定的参考价值。希望通过本书对大数据技术的推广和传播，让大数据技术走进我们的生活、学习和工作中。

由于作者水平有限，书中难免出现疏漏，敬请读者批评指正。

致谢

感谢人民邮电出版社责任编辑赵轩，因为他的辛勤工作才让本书的出版成为可能。

感谢曾经和我一起奋战在"大数据一线"的孟老师、马老师、游老师、赵老师、李老师。

最后，特别感谢我的父亲、母亲、岳父、岳母及我的妻子，是他们的全力支持才使我能够顺利完成此书。

杨力

2022 年 7 月

服务与支持

本书由异步社区出品,社区(https://www.epubit.com/)为您提供后续服务。

提交勘误信息

作者和编辑尽最大努力来确保书中内容的准确性,但难免会存在疏漏。欢迎您将发现的问题反馈给我们,帮助我们提升图书的质量。

当您发现错误时,请登录异步社区,按书名搜索,进入本书页面,单击"提交勘误",输入勘误信息,单击"提交"按钮即可,如下图所示。本书的作者和编辑会对您提交的勘误信息进行审核,确认并接受后,您将获赠异步社区的 100 积分。积分可用于在异步社区兑换优惠券、样书或奖品。

与我们联系

我们的联系邮箱是 contact@epubit.com.cn。

如果您对本书有任何疑问或建议,请您发邮件给我们,并请在邮件标题中注明本书书名,以便我们更高效地做出反馈。

如果您有兴趣出版图书、录制教学视频,或者参与图书翻译、技术审校等工作,可以发邮件给我们;有意出版图书的作者也可以到异步社区投稿(直接访问 www.epubit.com/

contribute 即可）。

如果您所在的学校、培训机构或企业想批量购买本书或异步社区出版的其他图书，也可以发邮件给我们。

如果您在网上发现有针对异步社区出品图书的各种形式的盗版行为，包括对图书全部或部分内容的非授权传播，请您将怀疑有侵权行为的链接通过邮件发送给我们。您的这一举动是对作者权益的保护，也是我们持续为您提供有价值的内容的动力之源。

关于异步社区和异步图书

"异步社区"是人民邮电出版社旗下IT专业图书社区，致力于出版精品IT图书和相关学习产品，为作译者提供优质出版服务。异步社区创办于2015年8月，提供大量精品IT图书和电子书，以及高品质技术文章和视频课程。

更多详情请访问异步社区官网 https://www.epubit.com。

"异步图书"是由异步社区编辑团队策划出版的精品IT专业图书的品牌，依托于人民邮电出版社的计算机图书出版积累和专业编辑团队，相关图书在封面上印有异步图书的LOGO。异步图书的出版领域包括软件开发、大数据、人工智能、测试、前端、网络技术等。

异步社区

微信服务号

目录

第1章 Scala 入门基础 ·············· 1
- 1.1 Scala 语言的特色 ················ 1
- 1.2 搭建 Scala 开发环境 ············· 3
  - 1.2.1 安装 JDK ···················· 3
  - 1.2.2 安装 Scala SDK ·············· 3
  - 1.2.3 安装 IDEA Scala 插件 ········ 4
- 1.3 Scala 解释器 ···················· 7
  - 1.3.1 启动 Scala 解释器 ············ 8
  - 1.3.2 执行 Scala 代码 ·············· 8
  - 1.3.3 退出 Scala 解释器 ············ 8
- 1.4 Scala 语法基础 ·················· 8
  - 1.4.1 定义变量 ···················· 8
  - 1.4.2 惰性赋值 ···················· 9
  - 1.4.3 字符串 ······················ 10
  - 1.4.4 数据类型与运算符 ·········· 11
  - 1.4.5 条件表达式 ·················· 12
- 1.5 Scala 控制结构和函数 ·········· 14
  - 1.5.1 for 表达式 ··················· 14
  - 1.5.2 while 循环 ··················· 16
  - 1.5.3 函数 ························· 16
  - 1.5.4 方法和函数的区别 ·········· 17
- 1.6 方法 ···························· 18
  - 1.6.1 定义方法 ···················· 18
  - 1.6.2 方法参数 ···················· 18
  - 1.6.3 方法调用方式 ················ 19
- 1.7 数组 ···························· 20
  - 1.7.1 定长数组 ···················· 21
  - 1.7.2 变长数组 ···················· 21
  - 1.7.3 遍历数组 ···················· 22
- 1.8 元组和列表 ······················ 23
  - 1.8.1 元组 ························· 23
  - 1.8.2 列表 ························· 24
  - 1.8.3 Set 集合 ····················· 30
- 1.9 Map 映射 ························ 32
  - 1.9.1 不可变 Map ·················· 32
  - 1.9.2 可变 Map ···················· 33
  - 1.9.3 Map 基本操作 ················ 33
- 1.10 函数式编程 ····················· 35
  - 1.10.1 遍历（foreach） ············ 35
  - 1.10.2 使用类型推断简化函数定义 ···· 36
  - 1.10.3 使用下画线简化函数定义 ······ 36
  - 1.10.4 映射（map） ················ 36
  - 1.10.5 扁平化映射（flatMap） ······ 37
  - 1.10.6 过滤（filter） ·············· 38
  - 1.10.7 排序 ························ 38
  - 1.10.8 分组（groupBy） ············ 40
  - 1.10.9 聚合（reduce） ············· 40
  - 1.10.10 折叠（fold） ··············· 41
- 1.11 本章总结 ······················· 42
- 1.12 本章习题 ······················· 42

第2章 Scala 面向对象编程 ······· 43
- 2.1 类与对象 ························ 43
- 2.2 定义和访问成员变量 ············ 44
- 2.3 使用下画线初始化成员变量 ····· 46

2.4　定义成员方法 ·············· 47
2.5　访问修饰符 ················ 48
2.6　类的构造器 ················ 51
　　2.6.1　主构造器 ············ 51
　　2.6.2　辅助构造器 ·········· 52
2.7　单例对象 ·················· 53
　　2.7.1　定义单例对象 ········ 53
　　2.7.2　在单例对象中定义成员方法 ········ 54
　　2.7.3　工具类案例 ·········· 54
2.8　main 方法 ················· 55
　　2.8.1　定义 main 方法 ······ 55
　　2.8.2　实现 App trait 来定义入口 ········· 55
2.9　伴生对象 ·················· 56
　　2.9.1　定义伴生对象 ········ 56
　　2.9.2　apply 和 unapply 方法 ········· 57
2.10　继承 ······················ 59
　　2.10.1　定义语法 ··········· 60
　　2.10.2　类继承 ············· 60
　　2.10.3　单例对象继承 ······· 61
　　2.10.4　override 和 super ··· 61
2.11　类型判断 ·················· 62
　　2.11.1　isInstanceOf 和 asInstanceOf 方法 ·············· 62
　　2.11.2　getClass 和 classOf ·· 63
2.12　抽象类 ···················· 64
2.13　匿名内部类 ················ 65
2.14　特质 ······················ 66
　　2.14.1　trait 作为接口使用 ··· 66
　　2.14.2　trait 中定义具体的字段和抽象字段 ············· 68
　　2.14.3　使用 trait 实现模板模式 ········ 69
　　2.14.4　对象混入 trait ······ 70
　　2.14.5　使用 trait 实现调用链模式 ········ 71
　　2.14.6　trait 调用链 ········ 72
　　2.14.7　trait 的构造机制 ···· 74
　　2.14.8　trait 继承类 ········ 74
2.15　本章总结 ·················· 75
2.16　本章习题 ·················· 75

第 3 章　Scala 编程高级应用 ······ 76
3.1　样例类 ···················· 76
　　3.1.1　定义样例类 ·········· 76
　　3.1.2　样例类方法 ·········· 77
　　3.1.3　样例对象 ············ 78
3.2　模式匹配 ·················· 79
　　3.2.1　简单匹配 ············ 79
　　3.2.2　守卫 ················ 80
　　3.2.3　匹配类型 ············ 80
　　3.2.4　匹配集合 ············ 81
　　3.2.5　变量声明中的模式匹配 ········ 82
　　3.2.6　匹配样例类 ·········· 83
3.3　Option 类型 ················ 83
3.4　偏函数 ···················· 84
3.5　正则表达式 ················ 85
3.6　异常处理 ·················· 86
　　3.6.1　捕获异常 ············ 86
　　3.6.2　抛出异常 ············ 87
3.7　提取器 ···················· 88
3.8　泛型 ······················ 89
　　3.8.1　定义泛型方法 ········ 90
　　3.8.2　定义泛型类 ·········· 90
　　3.8.3　上下界 ·············· 91
　　3.8.4　非变、协变和逆变 ···· 92
3.9　Actor ····················· 93
　　3.9.1　Java 并发编程的问题 ·· 94
　　3.9.2　Actor 并发编程模型 ··· 94
　　3.9.3　Java 并发编程与 Actor 并发编程 ··· 95
3.10　Actor 编程案例 ············ 95

3.10.1 创建 Actor …………………… 95
3.10.2 发送消息 / 接收消息 ………… 96
3.10.3 持续接收消息 ………………… 97
3.10.4 共享线程 …………………… 99
3.10.5 发送和接收自定义消息 ……… 99
3.10.6 基于 Actor 实现 WordCount 案例
　　　 …………………………… 101
3.11 本章总结 ………………………… 103
3.12 本章习题 ………………………… 103

第 4 章　Scala 函数式编程思想 …………… 104

4.1 作为值的函数 …………………… 104
4.2 匿名函数 ………………………… 105
4.3 柯里化 …………………………… 105
4.4 闭包 ……………………………… 106
4.5 隐式转换 ………………………… 107
4.6 隐式参数 ………………………… 109
4.7 Akka 并发编程框架 ……………… 109
　　4.7.1 Akka 特性 ………………… 110
　　4.7.2 Akka 通信过程 …………… 110
　　4.7.3 创建 ActorSystem ………… 111
4.8 Akka 编程入门案例 ……………… 111
　　4.8.1 实现步骤 ………………… 112
　　4.8.2 配置 Maven 模块依赖 …… 112
4.9 Akka 定时任务 …………………… 114
4.10 实现两个进程之间的通信 ……… 116
4.11 本章总结 ………………………… 119
4.12 本章习题 ………………………… 119

第 5 章　Spark 安装部署与入门 …… 120

5.1 Spark 简介 ……………………… 120
　　5.1.1 MapReduce 与 Spark ……… 120
　　5.1.2 Spark 组件 ………………… 122
　　5.1.3 Spark 生态系统 …………… 123

5.1.4 Spark 架构 ………………… 124
5.1.5 Spark 运行部署模式 ……… 125
5.1.6 Spark 远程过程调用协议 … 126
5.2 Spark 环境搭建 ………………… 126
　　5.2.1 本地模式部署 …………… 126
　　5.2.2 Standalone 集群模式 …… 128
　　5.2.3 Standalone-HA 集群模式 … 130
　　5.2.4 YARN 集群模式 ………… 132
　　5.2.5 Spark 命令 ………………… 137
5.3 编写 Spark 应用程序 …………… 139
　　5.3.1 Maven 简介 ……………… 140
　　5.3.2 安装 Maven ……………… 140
　　5.3.3 Spark 开发环境搭建 …… 141
　　5.3.4 配置 pom.xml 文件 ……… 143
　　5.3.5 开发应用程序——本地运行 … 147
　　5.3.6 修改应用程序——集群运行 … 148
　　5.3.7 集群硬件配置说明 ……… 150
5.4 本章总结 ………………………… 152
5.5 本章习题 ………………………… 152

第 6 章　SparkCore 编程 …… 153

6.1 RDD 概念与详解 ……………… 153
　　6.1.1 RDD 简介 ………………… 153
　　6.1.2 RDD 的主要属性 ………… 154
　　6.1.3 小结 ……………………… 156
6.2 RDD API 应用程序 ……………… 156
6.3 RDD 的方法（算子）分类 ……… 157
　　6.3.1 Transformation 算子 ……… 158
　　6.3.2 Action 算子 ……………… 160
6.4 基础练习 ………………………… 160
　　6.4.1 实现 WordCount 案例 …… 161
　　6.4.2 创建 RDD ………………… 162
　　6.4.3 map ……………………… 163
　　6.4.4 filter ……………………… 163
　　6.4.5 flatMap …………………… 164

6.4.6	sortBy	164
6.4.7	交集、并集、差集、笛卡尔积	165
6.4.8	groupByKey	165
6.4.9	groupBy	166
6.4.10	reduce	166
6.4.11	reduceByKey	166
6.4.12	repartition	167
6.4.13	count	167
6.4.14	top	168
6.4.15	take	168
6.4.16	first	168
6.4.17	keys、values	168
6.4.18	案例	169
6.5	实战案例	169
6.5.1	统计平均年龄	169
6.5.2	统计人口信息	171
6.5.3	在 IDEA 中实现 WordCount 案例	174
6.5.4	小结	176
6.6	RDD 持久化缓存	176
6.7	持久化缓存 API 详解	177
6.7.1	persist 方法和 cache 方法	177
6.7.2	存储级别	179
6.7.3	小结	181
6.8	RDD 容错机制 Checkpoint	181
6.8.1	代码演示	181
6.8.2	容错机制 Checkpoint 详解	182
6.9	本章总结	183
6.10	本章习题	183

第 7 章 SparkCore 运行原理 184

7.1	RDD 依赖关系	184
7.1.1	窄依赖与宽依赖	184
7.1.2	对比窄依赖与宽依赖	185
7.2	DAG 的生成和划分阶段	186

7.2.1	DAG 的生成	186
7.2.2	DAG 划分阶段	186
7.2.3	小结	188
7.3	Spark 原理初探	188
7.3.1	Spark 相关的应用概念	189
7.3.2	Spark 基本流程概述	191
7.3.3	流程图解	191
7.3.4	RDD 在 Spark 中的运行过程	192
7.3.5	小结	193
7.4	RDD 累加器和广播变量	193
7.4.1	累加器	194
7.4.2	广播变量	196
7.5	RDD 的数据源	198
7.5.1	普通文本文件	198
7.5.2	Hadoop API	199
7.5.3	SequenceFile	200
7.5.4	对象文件	201
7.5.5	HBase	202
7.5.6	JDBC	204
7.6	本章总结	207
7.7	本章习题	207

第 8 章 Spark SQL 结构化数据处理入门 208

8.1	数据分析方式	208
8.1.1	命令式	208
8.1.2	SQL 式	209
8.2	Spark SQL 的发展	210
8.3	数据分类和 Spark SQL 适用场景	211
8.3.1	结构化数据	211
8.3.2	半结构化数据	212
8.3.3	非结构化数据	213
8.4	Spark SQL 特点	214
8.5	Spark SQL 数据抽象	214
8.6	DataFrame 简介	215

8.7　Dataset 简介 ·············· 215
8.8　RDD、DataFrame 和 Dataset 的
　　区别 ····················· 216
8.9　Spark SQL 初体验 ·········· 218
　　8.9.1　SparkSession 入口 ····· 218
　　8.9.2　创建 DataFrame ······· 219
　　8.9.3　创建 Dataset ·········· 222
　　8.9.4　两种查询风格 ········· 224
8.10　本章总结 ················ 229
8.11　本章习题 ················ 229

第 9 章　Spark SQL 结构化数据
　　　　处理高级应用 ········ 230
9.1　使用 IDEA 开发 Spark SQL ···· 230
　　9.1.1　创建 DataFrame 和 Dataset ·· 231
　　9.1.2　花式查询 ············ 233
　　9.1.3　相互转换 ············ 235
　　9.1.4　Spark SQL 词频统计实战 ·· 237
9.2　Spark SQL 多数据源交互 ····· 240
9.3　Spark SQL 自定义函数 ······· 242
　　9.3.1　自定义函数分类 ······· 242
　　9.3.2　UDF ················ 243
　　9.3.3　UDAF ··············· 244
9.4　Spark on Hive ············· 247
　　9.4.1　开启 Hive 的元数据库服务 ·· 247
　　9.4.2　Spark SQL 整合 Hive
　　　　元数据库 ············ 248
　　9.4.3　使用 Spark SQL 操作 Hive 表 ·· 248
9.5　本章总结 ················· 249
9.6　本章习题 ················· 249

第 10 章　Spark Streaming 核心编程
　　　　　 ···················· 250
10.1　场景需求 ················ 250
10.2　Spark Streaming 概述 ······· 251

10.2.1　Spark Streaming 的特点 ····· 252
10.2.2　Spark Streaming 实时计算
　　　所处的位置 ··········· 252
10.3　Spark Streaming 原理 ········ 254
　　10.3.1　基本流程 ············ 255
　　10.3.2　数据模型 ············ 255
10.4　DStream 相关的 API ········ 256
　　10.4.1　Transformation ········· 257
　　10.4.2　Output ·············· 257
10.5　Spark Streaming 原理总结 ···· 258
10.6　Spark Streaming 实战 ······· 258
　　Spark Streaming 第一个案例
　　WordCount ················ 258
10.7　updateStateByKey 算子 ······ 262
　　10.7.1　WordCount 案例问题分析 ·· 262
　　10.7.2　代码实现 ············ 262
　　10.7.3　执行步骤 ············ 263
10.8　reduceByKeyAndWindow 算子 · 264
　　10.8.1　图解 reduceByKeyAndWindow
　　　　算子 ················ 264
　　10.8.2　代码实现 ············ 264
　　10.8.3　执行步骤 ············ 266
10.9　统计一定时间内的热搜词 ···· 266
　　10.9.1　需求分析 ············ 266
　　10.9.2　代码实现 ············ 266
　　10.9.3　执行步骤 ············ 268
10.10　整合 Kafka ·············· 268
　　10.10.1　Kafka 基本概念 ······· 268
　　10.10.2　Kafka 的特性 ········· 268
　　10.10.3　核心概念图解 ········ 269
　　10.10.4　Kafka 集群部署 ······· 270
　　10.10.5　Kafka 常用命令 ······· 275
　　10.10.6　Receiver 接收方式 ····· 276
　　10.10.7　Direct 直连方式 ······· 277

10.10.8 spark-streaming-kafka-0-8 版本 ·············· 277
10.10.9 spark-streaming-kafka-0-10 版本 ·············· 282
10.11 本章总结 ·············· 285
10.12 本章习题 ·············· 285

第 11 章　Spark 综合项目实战 ··· 286

11.1 网站运营指标统计项目 ·············· 286
  11.1.1 需求分析 ·············· 286
  11.1.2 数据分析 ·············· 287
  11.1.3 代码实现 ·············· 287
11.2 热力图分析项目 ·············· 289
  11.2.1 需求分析 ·············· 289
  11.2.2 数据分析 ·············· 289
  11.2.3 项目开发 ·············· 290
11.3 本章总结 ·············· 300
11.4 本章习题 ·············· 300

第 1 章 Scala 入门基础

Scala 是面向对象编程和函数式编程这两种编程范式相结合的编程语言，也是一门静态类型的编程语言。Scala 程序的执行需要编译和解释，其运行环境就是 JVM（Java Virtual Machine，Java 虚拟机）。Scala 语言的静态类型特性有助于开发者避免在编写的复杂应用程序中出现漏洞（bug）。Scala 的 JVM 运行环境和类 JavaScript 动态运行时特性，允许开发者去构建高性能应用系统，并且可以轻松访问巨大的 Java 类库和 Scala 库生态系统。

1.1 Scala 语言的特色

Scala 语言最初并没有引起开发者的重视，但随着"大数据时代"的到来，人们发现两个大数据处理框架 Spark 和 Kafka 竟都是用 Scala 语言开发出来的。至此，Scala 这门"沉睡"已久的语言逐步进入大数据开发者的眼帘。

Scala 语言表达能力强，能用少量代码实现需要用大量 Java 代码才能实现的功能。比如，我们要创建一个学生类，属性包括学生学号、学生姓名、学生年龄、班级信息。要实现这个功能，Java 代码如下。

```
public class Student {                      //学生类
    private int studentNum;                 //学生学号
    private String studentName;             //学生姓名
    private int studentAge;                 //学生年龄
    private String classInfo;               //班级信息
```

```java
    public int getStudentNum() {
        return studentNum;
    }

    public void setStudentNum(int studentNum) {
        this.studentNum = studentNum;
    }

    public String getStudentName() {
        return studentName;
    }

    public void setStudentName(String studentName) {
        this.studentName = studentName;
    }

    public int getStudentAge() {
        return studentAge;
    }

    public void setStudentAge(int studentAge) {
        this.studentAge = studentAge;
    }

    public String getClassInfo() {
        return classInfo;
    }

    public void setClassInfo(String classInfo) {
        this.classInfo = classInfo;
    }

    /**
     * 重写toString 方法
     * @return String
     */
    @Override
    public String toString() {
        return "Student{" +
                "studentNum=" + studentNum +
                ", studentName='" + studentName + '\'' +
```

```
              ", studentAge=" + studentAge +
              ", classInfo='" + classInfo + '\'' +
              '}';
    }
}
```

要实现与上面 Java 代码相同的功能,Scala 代码如下。

```
case class Student( var studentNum:int,  var studentName:String,  var stu-
dentAge:int,  var classInfo:String)      //学生类
```

可以看到,Scala 代码更为简洁,可以极大地提高开发效率。此外,Scala 兼容 Java,可以访问庞大的 Java 类库。我们在 Scala 中调用 Java 程序的 Scanner 类并进行对话,代码如下。

```
scala> import java.util.Scanner
import java.util.Scanner

// 光标出现后输入 hello scala
scala> println("hello world :" + new Scanner(System.in).nextLine())
hello world :hello scala
```

1.2 搭建 Scala 开发环境

Scala 开发环境搭建分为两步,首先安装 JDK(Java Development Kit,Java 语言开发工具包),然后安装 SDK(Scala Development Kit,Scala 语言开发工具包)。

1.2.1 安装 JDK

Scala 程序的运行环境为 JVM,其中经常会用到 Java 类库,所以我们在使用 Scala 前,一定要先安装 JDK(1.8 及以上版本),并配置好环境变量。

1.2.2 安装 Scala SDK

安装好 JDK 后,还需要安装 Scala SDK 才可编译、运行 Scala 代码。在 Scala 官方网站下载适合的版本,如图 1-1 所示。

Archive	System	Size
scala-2.11.12.tgz	Mac OS X, Unix, Cygwin	27.77M
scala-2.11.12.msi	Windows (msi installer)	109.82M
scala-2.11.12.zip	Windows	27.82M
scala-2.11.12.deb	Debian	76.44M
scala-2.11.12.rpm	RPM package	108.60M
scala-docs-2.11.12.txz	API docs	46.17M
scala-docs-2.11.12.zip	API docs	84.26M
scala-sources-2.11.12.tar.gz	Sources	

图 1-1

以下是 Scala SDK 的安装流程（以 Linux 版本为例）。

① 在 Scala 官方网站下载名为 scala-2.11.12.tgz 的文件，并将其放到 /opt/scala 目录下。

② 进入 /opt/scala 目录，解压并安装 scala-2.11.12.tgz 文件，命令如下。

```
cd /opt/scala
tar -zxvf scala-2.11.12.tgz
```

③ 打开 profile 文件，配置环境变量，命令如下。

```
vi /etc/profile
export SCALA_HOME=/opt/scala/scala-2.11.12
export PATH=${SCALA_HOME}/bin:${PATH}
```

④ 保存并关闭 profile 文件，之后执行下面的命令，使 profile 文件重新生效。

```
source /etc/profile
```

⑤ 验证安装是否成功，命令如下。

```
scalac -version
```

1.2.3　安装 IDEA Scala 插件

IDEA 的全称是 IntelliJ IDEA，是 Java IDE（Integrated Development Environment，集成开发环境），但其默认不支持 Scala 语言，为此，我们需要下载和配置 Scala 插件。

① 登录 IDEA 官方网站，下载并安装最新版本的 IDEA 开发工具，如图 1-2 所示。

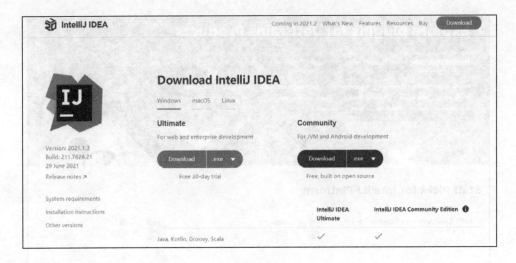

图 1-2

② 在安装向导中选中 Plugins，在搜索框中输入 scala，单击 Install 按钮，如图 1-3 所示。

图 1-3

如果在下载过程中遇到问题，也可以到 IDEA 官方网站下载 Scala 插件，如图 1-4 所示。

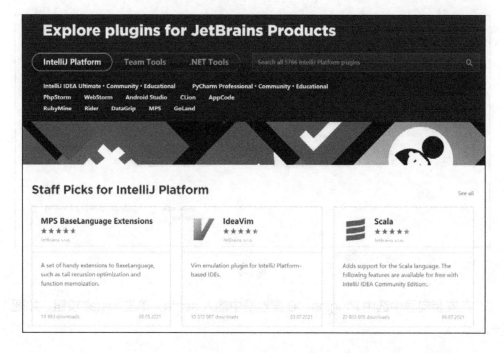

图 1-4

③ 选中 Scala 选项并单击 Get 按钮后，可以看到 Scala 下载界面，其中第一列代表 Scala 版本，第二列代表 IDEA 版本。由于当前我们使用的 IDEA 为 2021.1.3 版本，因此下载 2021.1.21 版本的 Scala 插件，如图 1-5 所示。

图 1-5

④ 将下载好的 jar 包通过图 1-6 所示的形式导入，重启 IDEA 即可。

图 1-6

⑤ 如果出现图 1-7 所示的界面则表示安装成功，现在我们就可以在 IDEA 中编写 Scala 代码了。

图 1-7

1.3 Scala 解释器

Scala SDK 中提供了 Scala 解释器，能帮助我们更方便地执行 Scala 程序。Scala 解释器经常用于代码测试。

1.3.1 启动 Scala 解释器

在命令行中执行 scala 命令，就可以启动 Scala 解释器，如图 1-8 所示。

```
Welcome to Scala 2.11.12 (OpenJDK 64-Bit Server VM, Java 1.8.0_262).
Type in expressions for evaluation. Or try :help.

scala>
```

图 1-8

1.3.2 执行 Scala 代码

在命令行界面执行 print("hello world") 命令，如图 1-9 所示。

```
Welcome to Scala 2.11.12 (OpenJDK 64-Bit Server VM, Java 1.8.0_262).
Type in expressions for evaluation. Or try :help.

scala> print("hello world")
hello world
```

图 1-9

1.3.3 退出 Scala 解释器

在命令行执行 :quit 命令或者按快捷键 Ctrl+D 即可退出 Scala 解释器。

1.4 Scala 语法基础

想要熟练掌握 Scala 语言，不可忽视 Scala 语言语法的学习。

1.4.1 定义变量

Scala 的变量是如何定义的呢？

比如，我们在 Java 中以"变量类型 变量名 = 初始值"的方式定义学生年龄。

```
int studentAge = 18;
```

而在 Scala 中，我们需要通过不同的方式来确定变量是否不可变（只读）：val/var 变量名 : 变量类型 = 初始值。

变量类型可以指定，也可以不指定，由初始化的值决定。这种由初始化的值决定变量类型的方式叫作类型推断。var 代表可变变量，val 代表不可变变量。

使用定义格式定义变量的方法如下。

```
scala> var studentAge:Int =18
studentAge: Int = 18
```

使用类型推断的形式定义变量的方法如下。

```
scala> var studentAge =18
studentAge: Int = 18
```

那么 var 和 val 在什么地方不一样呢？现在使用 var 定义变量并修改值。

```
scala> var studentAge =18
studentAge: Int = 18
scala> studentAge = 19
studentAge: Int = 19
```

可以看到，使用 var 定义的学生年龄可以被修改。现在使用 val 定义变量并修改值。

```
scala> val studentAge =18
studentAge: Int = 18
scala> studentAge = 19
<console>:12: error: reassignment to val
       studentAge=19
                  ^
```

在修改使用 val 定义的学生年龄时报错，说明使用 val 定义的变量是只读的，不可修改。

1.4.2 惰性赋值

使用 var 或者 val 定义的变量会直接加载到 JVM 内存，但在一些大型计算场景下，如执行一条复杂 SQL 语句，这个 SQL 语句可能有成百上千行，直接加载到 JVM 内存可能会造成很大的内存开销。针对这种情形，Scala 提供了一个更好的赋值方式，也就是惰性赋值。当一个变量需要保存很大的数据，且不想直接加载到 JVM 内存，而是等需要执行时再加载时可以使用惰性赋值。其语法格式为：lazy val/var 变量名 = 表达式。

```
scala> lazy val sql = """
| select a.student_name ,b.class_info, c.teacher_name
| from a ,b,c
| where a.student_name in b.student_name
| ...     // 此为上百行查询条件
| group by a.class_score """
sql: String = <lazy>
```

1.4.3 字符串

Scala 提供了多种字符串的定义方式，我们可以根据需要选择。具体定义方式和使用场景及其案例如下。

1. 双引号

最简单的双引号，适用于大多数字符串定义，如定义一个学生的名称，值为 Lisa。

```
scala> val studentName = "Lisa"
studentName: String = Lisa
```

2. 插值表达式

插值表达式定义的语法格式为 val/var 变量名 = s"${变量/表达式}字符串"，适用于需要进行字符串拼接的场景，如定义一场考试的试卷信息，具体试卷类型由试卷编号和年级决定。

```
scala> val examNum = "001"
examNum: String = 001
scala> val classInfo = "classNo.1"
classInfo: String = classNo.1
scala> val examInfo = s"${classInfo}+${examNum}"
examInfo: String = classNo.1+001
```

3. 三引号

三引号定义的语法格式为 val 变量名 = """""""，适用于有大段文字需要保存的场景，希望可以换行且不影响数据定义，方便阅读，如定义一个 SQL 语句。

```
scala> val sql = """
     | select student_name, class_score, class_info
     | from score_table
     | where class_score > 90 """
sql: String =
" select student_name, class_score, class_info
from score_table
where class_score > 90 "
```

1.4.4 数据类型与运算符

Scala 语言像其他编程语言一样有数据类型和运算符。

1. 数据类型

Scala 的数据类型和 Java 的大同小异。需要注意的是，Scala 中所有数据类型的首字母都是大写的，整数类型是 Int，而不是 Integer（见表 1-1）。

表 1-1

数据类型	描述
Byte	8 位有符号补码整数，数值区间为 -128 到 127
Short	16 位有符号补码整数，数值区间为 -32768 到 32767
Int	32 位有符号补码整数，数值区间为 -2147483648 到 2147483647
Long	64 位有符号补码整数，数值区间为 -9223372036854775808 到 9223372036854775807
Float	32 位 IEEE 754 标准的单精度浮点数
Double	64 位 IEEE 754 标准的双精度浮点数
Char	16 位无符号 Unicode 字符，区间值为 U+0000 到 U+FFFF
String	字符序列
Boolean	值为 true 或 false

2. 运算符

Scala 提供了丰富的内置运算符（见表 1-2），主要作用是告诉编译器执行特定的数学或逻辑函数运算。

表1-2

类别	运算符
算术运算符	+、-、*、/、%
关系运算符	==、!=、>、<、>=、<=
逻辑运算符	&&、||、!
按位运算符	&、|、^

注意，Scala 中没有 ++ 和 -- 这两种运算符。要比较两个值是否相等，直接使用 == 或者 != 即可，它们相当于 Java 中的 equals。

```
scala> var a =0
a: Int = 0
scala> a++
<console>:13: error: value ++ is not a member of Int
a++
^
scala> var b =0
a: Int = 0
scala> b--
<console>:13: error: value -- is not a member of Int
b--
^
```

3. Scala 类型层次结构

Any 是所有类型的父类，定义了一些通用的方法，如 toString、equals、hashCode 等。Any 有两个子类：AnyVal 是所有数值类型的父类，AnyRef 是所有对象类型（引用类型）的父类。Nothing 是所有类型的子类型，但是没有一个值是 Nothing 类型的，它的主要作用是给出非正常终止的信号，比如抛出异常、程序退出等。Null 是所有引用类型的子类型，即 AnyRef 的子类，它由值 null 来表示，多用于和其他语言的互操作。

1.4.5 条件表达式

条件表达式就是 if 表达式，根据条件执行对应的命令。

1. 有返回值的 if 表达式

Java 中的 if 表达式是没有返回值的,如定义变量 a =1、变量 b=2,如果变量 a 等于变量 b,则返回 true,否则返回 false。

```
scala> val a = 1
a: Int = 1

scala> val b = 2
b: Int = 2

scala> if (a == b) "true" else "false"
res2: String = false
```

可以看出,if 表达式的返回值为 false。在 Java 中,需要再定义一个变量去赋值,才能得到命令执行后的值。

Scala 没有为三元运算设计特定的运算符,我们可以利用 if 表达式带有返回值的特性来模拟。

```
scala> val c = if (a == b) "true" else "false"
c: String = false
```

2. 块表达式

块表达式用于多行处理语句赋值一个变量,多行处理语句用 {} 包裹后作用于一个变量。

```
scala> val c = {
     | print("c result is")
     |   if (a == b) "true" else "false"
     | }
c result is c: String = false
```

注意,上面示例中的 if 判断语句,其返回 false,则 c 的值为 false。那么,如果我们将 print 放在最后一行会有什么变化呢?

```
scala> val c  = {
     | if (a == b) "true" else "false"
     | print("c result is ")
     | }
c result is c: Unit = ()
```

c 的值变为 null,这是因为 print 的返回值为 null。

1.5 Scala 控制结构和函数

Scala 语言提供了基本的控制结构和丰富的函数。

1.5.1 for 表达式

for 表达式的语法格式如下。

```
for( var x to Range ){
    // 表达式
}
for( var x <- Range ){
    // 表达式
}
```

其中，to 代表包含数组 length。

1. 简单循环

我们可以通过 .to 的方式使用 for 循环表达式输出数字 1 ~ 10。

```
scala> val fun1 = 1.to(10)      // 使用函数.to生成数组
fun1: scala.collection.immutable.Range.Inclusive = Range(1, 2, 3, 4, 5, 6, 7, 8, 9, 10)

scala> for(i <-fun1) println(i)
1
2
3
4
5
6
7
8
9
10
```

我们也可以在 for 循环中直接使用 to 输出值。

```
scala> for(i <- 1 to 10) println(i)
1
2
3
4
5
6
7
8
9
10
```

2. 嵌套循环

乘法表是嵌套循环的简单示例。

```
scala> for(i<-1 to 9;j<-1 to i){
     |     print(j+"*"+i+"="+j*i+" ")
     |     if(j==i)  println()
     | }
1*1=1
1*2=2 2*2=4
1*3=3 2*3=6  3*3=9
1*4=4 2*4=8  3*4=12 4*4=16
1*5=5 2*5=10 3*5=15 4*5=20 5*5=25
1*6=6 2*6=12 3*6=18 4*6=24 5*6=30 6*6=36
1*7=7 2*7=14 3*7=21 4*7=28 5*7=35 6*7=42 7*7=49
1*8=8 2*8=16 3*8=24 4*8=32 5*8=40 6*8=48 7*8=56 8*8=64
1*9=9 2*9=18 3*9=27 4*9=36 5*9=45 6*9=54 7*9=63 8*9=72 9*9=81
```

3. 守卫

在 for 循环中添加 if 判断语句作为过滤条件，if 判断语句称为守卫。

```
for(i <- 表达式/数组/集合 if 表达式) {
// 表达式
}
```

比如，循环输出数字1～3，并且过滤掉数字2。

```
scala> for(i <- 1 to 3 if i != 2) {
     |     print(i + " ")
     | }
1 3
```

4. for 推导式

for 推导式主要用来将 for 循环后的数据返回到一个变量，它区别于 Java 代码用对象承接的方式，而是使用 yield 直接将数据返回。

```
val v = for(i <- Range ) yield 表达式
```

比如，循环打印数字 1 ~ 3，并且过滤掉数字 2，最后返回到一个变量。

```
scala> val res = for(i <- 1 to 3 if i != 2) yield {
     |     i
     | }
res: scala.collection.immutable.IndexedSeq[Int] = Vector(1, 3)
```

1.5.2　while 循环

对于 while 循环，Scala 和 Java 是一样的，比如打印 1 ~ 10 的数字。

```
scala> var i = 1
i: Int = 1

scala>   while(i <= 10){
     |      println(i)
     |      i+=1
     | }
1
2
3
4
5
6
7
8
9
10
```

1.5.3　函数

在 Scala 中，我们可以像操作数字一样将函数赋值给一个变量。

使用 val 语句定义函数的方式为：val 函数名 = ([参数名 : 参数类型]) => 函数体。

定义两个数相加的函数如下。

```
scala> val add = (a:Int, b:Int) => a + b
add: (Int, Int) => Int = <function2>

scala> add(1,2)
res2: Int = 3
```

1.5.4 方法和函数的区别

与函数不同，方法需要使用 def 来定义：def 方法名 (参数名：参数类型，参数名：参数类型):[返回类型] => { 方法体 }。

函数是对象（变量），而方法不是。方法在运行时，加载到 JVM 方法区；函数在运行时，加载到 JVM 堆内存。

函数是对象，继承 FunctionN，函数对象有 apply、curried、toString、tupled 这些方法，方法则没有。方法无法像函数一样赋值给变量。

```
scala> val f = fun1
<console>:12: error: missing argument list for method fun1
Unapplied methods are only converted to functions when a function type is expected.
You can make this conversion explicit by writing 'fun1 _' or 'fun1(_,_)' instead of 'fun1'.
       val f = fun1
               ^
```

想要将方法赋值给一个对象，可以先将方法转换为函数后再进行赋值。

将方法转换成函数的语法格式为 val 函数名 = 方法名 _。

```
scala> val f = fun1 _
f: (Int, Int) => Int = <function2>

scala> def fun1(a:Int,b:Int)= a+b
fun1: (a: Int, b: Int)Int

scala> val f = fun1 _
f: (Int, Int) => Int = <function2>
```

1.6 方法

在 Scala 中可以定义多个方法，Scala 中的方法和 Java 类似，但是定义方式不同。

1.6.1 定义方法

定义方法的语法格式为：def 方法名 (参数名 : 参数类型 , 参数名 : 参数类型):[返回类型] => { 方法体 }。

需要注意的是，参数列表中的参数类型不能省略，返回类型可省略，最后方法类型由 Scala 自动推断。当方法体中没有 return 明确返回值时，由块表达式的值决定返回值。比如，定义两个数相加的方法，并求出 1+2 的代码如下。

```
scala> def add(x:Int, y:Int):Int = x + y
add: (x: Int, y: Int)Int

scala> add(1,2)
res5: Int = 3
```

1.6.2 方法参数

方法参数主要有默认参数、带名参数和变长参数这 3 种。

1. 默认参数

在方法定义参数的时候，给参数一个默认值，当调用该方法时，如果没有传入对应的参数值，那么会使用这个默认值进行处理。

比如，定义一个方法，让两个数相加，默认第一个参数值为 3，第二个参数值为 4，调用方法时不进行传值，求最后返回的值是多少。

```
scala> def add(x:Int=3, y:Int=4):Int = x + y
add: (x: Int, y: Int)Int

scala> add()
res6: Int = 7
```

2. 带名参数

在调用方法时，指定参数名称进行传值。

比如，定义一个方法，让两个数相加，默认第一个参数值为 3，第二个参数值为 4，调用方法时指定第一个参数值为 5，求最后返回的值是多少。

```
scala> def add(x:Int=3, y:Int=4):Int = x + y
add: (x: Int, y: Int)Int

scala> add(x=5)
res7: Int = 9
```

3. 变长参数

在实际开发中，如果方法的参数个数不固定，那么可以使用变长参数来定义，在参数后面加"*"，表示参数可能是 0 个，也可能是多个，语法格式为：def 方法名 (参数名：参数类型 *): 返回类型 = { 方法体 }。比如，定义一个多个数字相加的方法。

```
scala>  def add(n:Int*) = n.sum
add: (n: Int*)Int

scala> add(1,2,3,4,5,6,7,8,9,10)
res8: Int = 55
```

1.6.3 方法调用方式

在 Scala 中，有多种调用方法的方式：后缀调用法、中缀调用法、花括号调用法、无括号调用法。

1. 后缀调用法

和在 Java 中调用方法一样，调用方式为：对象名 . 方法名 (参数)。比如，计算 3.14 向上取整的代码如下。

```
scala> Math.ceil(3.14)
res11: Double = 4.0
```

2. 中缀调用法

Scala 中的所有运算符都包含方法，例如加号，它底层就是一个方法。中缀调用法的调用方式为：对象名 方法名 参数（如果有多个参数，参数部分用圆括号包裹）。比如，计算 3.14 向上取整的代码如下。

```
scala> Math ceil 3.14
res12: Double = 4.0
```

3. 花括号调用法

只有参数的个数为 1 时，才可以使用花括号（{}）调用，调用方式为：对象名 . 方法名 { 参数 }。比如，计算 3.14 向上取整的代码如下。

```
scala> Math.ceil{3.14}
res13: Double = 4.0
```

当传入多个参数时会报错。

```
scala> Math.ceil{3.14 , 3.14}
<console>:1: error: ';' expected but ',' found.
Math.ceil{3.14 , 3.14}
```

4. 无括号调用法

当定义的方法没有参数时，可以省略括号进行调用。

```
scala> def fun1()= {1+1}
fun1: ()Int
scala> fun1
res1: Int = 2
```

1.7 数组

和 Java 类似，Scala 也提供了存储一组数据的容器，主要有定长数组和变长数组。Scala 数组需要使用 () 取值，而且 Scala 数组中可以存放不同类型的数据，最终数据的类型是其共同父类。

比如，定义一个数组，里面包含 Int、String 和 Boolean 类型的值。

```
scala> val a = Array(1,"2",true)
a: Array[Any] = Array(1, 2, true)
```

1.7.1 定长数组

定长数组，顾名思义，指的是数组长度是不可变的。定义定长数组的方法有如下两种。

方法 1：val/var 变量名 = **new** Array[元素类型](数组长度)   // 通过指定长度定义

比如，定义一个长度为 10 的 Int 类型数组。

```
scala> val a = new Array[Int](10)
a: Array[Int] = Array(0, 0, 0, 0, 0, 0, 0, 0, 0, 0)
```

方法 2：val/var 变量名 = Array(元素1, 元素2, 元素3, ...)   // 使用元素直接定义

比如，定义一个数组元素为 1、2、3 的数组。

```
scala> val a = Array(1,2,3)
a: Array[Int] = Array(1, 2, 3)
```

1.7.2 变长数组

变长数组是指数组元素的个数可以增减，即我们可以为其添加、删除、修改元素。最重要的一点是，要提前导入 ArrayBuffer 类：import scala.collection.mutable.ArrayBuffer。

定义空变长数组的语法格式为：val/var a = ArrayBuffer[元素类型]()。

定义带有初始元素的变长数组的语法格式为：val/var a = ArrayBuffer(元素1, 元素2, 元素3, ...)。比如，定义一个变长数组，初始元素为 1、2、3。

```
scala> import scala.collection.mutable.ArrayBuffer
import scala.collection.mutable.ArrayBuffer

scala> val a = ArrayBuffer(1, 2, 3)
a: scala.collection.mutable.ArrayBuffer[Int] = ArrayBuffer(1, 2, 3)
```

添加元素到变长数组的方式为：+=。比如，向变长数组 a 中添加元素 4。

```
scala> import scala.collection.mutable.ArrayBuffer
import scala.collection.mutable.ArrayBuffer
```

```
scala> val a = ArrayBuffer(1, 2, 3)
a: scala.collection.mutable.ArrayBuffer[Int] = ArrayBuffer(1, 2, 3)

scala> a+=4
res0: a.type = ArrayBuffer(1, 2, 3, 4)
```

添加另一个数组到此变长数组的方式为：++=。比如，添加一个数组到变长数组。

```
scala> import scala.collection.mutable.ArrayBuffer
import scala.collection.mutable.ArrayBuffer

scala> val a = ArrayBuffer(1, 2, 3, 4)
a: scala.collection.mutable.ArrayBuffer[Int] = ArrayBuffer(1, 2, 3, 4)

scala> a++=Array(5,6)
res2: a.type = ArrayBuffer(1, 2, 3, 4, 5, 6)
```

删除变长数组中的一个元素的方式为：-=。比如，删除变长数组 a 中值等于 6 的元素。

```
scala> import scala.collection.mutable.ArrayBuffer
import scala.collection.mutable.ArrayBuffer

scala> val a = ArrayBuffer(1, 2, 3, 4, 5 ,6)
a: scala.collection.mutable.ArrayBuffer[Int] = ArrayBuffer(1, 2, 3, 4, 5, 6)
scala> a-=6
res3: a.type = ArrayBuffer(1, 2, 3, 4, 5)
```

1.7.3 遍历数组

和在 Java 中一样，在 Scala 中可以通过索引或者 for 表达式直接遍历数组。

```
scala> val a = Array(1,2,3)
a: Array[Int] = Array(1, 2, 3)

scala> for(i<-a) println(i)
```

通过索引形式遍历数组的方式如下。

```
scala> val a = Array(1,2,3)
a: Array[Int] = Array(1, 2, 3)

scala> for(i <- 0 until a.length) println(a(i))
```

1.8 元组和列表

1.8.1 元组

元组可以存放不同数据类型的数据，且元组中的元素是不可变的。比如，定义一个元组，其包含数字 1、字符串 "scala" 和布尔值 true，若试图改变数字 1 为数字 3，会报错。

```
scala> val a =(1,"scala",true)
a: (Int, String, Boolean) = (1,scala,true)

scala> a._1
res0: Int = 1

scala> a._1 = 3
<console>:12: error: reassignment to val
       a._1 = 3
         ^
```

1. 定义元组

定义元组有两种方式：一是使用 () 定义，二是使用 -> 定义。使用 -> 定义的形式只适用于两个元素。推荐使用 () 定义。

示例：定义一个元组，其包含数字 1、字符串 "scala" 和布尔值 true。

```
scala> val a =(1,"scala",true)
a: (Int, String, Boolean) = (1,scala,true)
scala> val a = 1->2
a: (Int, Int) = (1,2)
scala> val a = 1->2,3->4
<console>:1: error: ';' expected but ',' found.
val a = 1->2,3->4
            ^
```

2. 访问元组

访问元组和访问数组不同，访问元组需要使用 ._index，index 从 1 开始。使用 ._1

访问的是元组中的第一个元素，使用 ._2 访问的是元组中的第二个元素。

示例：定义一个元组，其包含数字 1、字符串 "scala" 和布尔值 true，并一次访问其中的元素。

```
scala> val a =(1,"scala",true)
a: (Int, String, Boolean) = (1,scala,true)

scala> a._1
res1: Int = 1

scala> a._2
res2: String = scala

scala> a._3
res3: Boolean = true
```

1.8.2 列表

列表是 Scala 中使用非常多的数据结构，其特点是可以保存重复的值，并且有先后顺序。列表和数组一样，也分为可变和不可变两种。使用可变列表前要先使用命令 import scala.collection.mutable.ListBuffer。

1. 不可变列表

不可变列表就是其元素和长度都不可更改。

创建不可变列表的语法格式如下。

```
val/var 变量名 = List(元素1，元素2，元素3,...)    //创建带有初始元素的不可变列表
val/var 变量名 = 元素1 :: 元素2 :: Nil           //创建带有初始元素的不可变列表
val/var 变量名 = Nil     // 创建一个空的不可变列表
```

根据上述 3 种创建方式，创建一个元素为 0、1、2、3 的列表。

```
scala> val list = List(0,1,2,3)
list: List[Int] = List(0, 1, 2, 3)

scala> val list = 0::1::2::3::Nil
list: List[Int] = List(0, 1, 2, 3)
```

```
scala> val list = Nil
list: scala.collection.immutable.Nil.type = List()
```

2. 可变列表

可变列表就是其元素和长度都可以改变。

创建可变列表的语法格式如下。

```
// 创建空的可变列表
val/var 变量名 = ListBuffer[Int]()
// 创建带有初始元素的可变列表
val/var 变量名 = ListBuffer(元素1,元素2,元素3,...)
```

比如，创建一个空的可变列表。

```
scala> import scala.collection.mutable.ListBuffer
import scala.collection.mutable.ListBuffer

scala> val list = ListBuffer()
list: scala.collection.mutable.ListBuffer[Nothing] = ListBuffer()
```

比如，创建一个元素为0、1、2、3的可变列表。

```
scala> import scala.collection.mutable.ListBuffer
import scala.collection.mutable.ListBuffer

scala> val list = ListBuffer(0,1,2,3)
list: scala.collection.mutable.ListBuffer[Int] = ListBuffer(0, 1, 2, 3)
```

对列表的常见操作是增删改查元素、将可变列表转变为不可变的列表和数组。增删改查操作方式和可变数组类似。

示例：定义一个可变列表，初始元素为0、1、2、3，获取第一个元素。

```
scala> import scala.collection.mutable.ListBuffer
import scala.collection.mutable.ListBuffer

scala> val list = ListBuffer(0,1,2,3)
list: scala.collection.mutable.ListBuffer[Int] = ListBuffer(0, 1, 2, 3)

scala> list(0)
res0: Int = 0
```

示例：定义一个可变列表，初始元素为0、1、2、3，添加一个元素4到尾部。

```
scala> import scala.collection.mutable.ListBuffer
import scala.collection.mutable.ListBuffer

scala> val list = ListBuffer(0,1,2,3)
list: scala.collection.mutable.ListBuffer[Int] = ListBuffer(0, 1, 2, 3)

scala> list +=4
res1: list.type = ListBuffer(0, 1, 2, 3, 4)
```

示例：定义一个可变列表，初始元素为 0、1、2、3，删除元素 3。

```
scala> import scala.collection.mutable.ListBuffer
import scala.collection.mutable.ListBuffer

scala> val list = ListBuffer(0,1,2,3)
list: scala.collection.mutable.ListBuffer[Int] = ListBuffer(0, 1, 2, 3)

scala> list -=3
res2: list.type = ListBuffer(0, 1, 2)
```

示例：定义一个可变列表，初始元素为 0、1、2、3，追加一个新的列表。

```
scala> import scala.collection.mutable.ListBuffer
import scala.collection.mutable.ListBuffer

scala> val list = ListBuffer(0,1,2,3)
list: scala.collection.mutable.ListBuffer[Int] = ListBuffer(0, 1, 2, 3)

scala> list++=List(4,5)
res3: list.type = ListBuffer(0, 1, 2, 3, 4, 5)
```

示例：定义一个可变列表，初始元素为 0、1、2、3，将可变列表转变为不可变的列表。

```
scala> import scala.collection.mutable.ListBuffer
import scala.collection.mutable.ListBuffer

scala> val list = ListBuffer(0,1,2,3)
list: scala.collection.mutable.ListBuffer[Int] = ListBuffer(0, 1, 2, 3)

scala> list.toList
res4: List[Int] = List(0, 1, 2, 3)
```

示例：定义一个可变列表，初始元素为 0、1、2、3，将可变列表转变为不可变的数组。

```
scala> import scala.collection.mutable.ListBuffer
import scala.collection.mutable.ListBuffer

scala> val list = ListBuffer(0,1,2,3)
list: scala.collection.mutable.ListBuffer[Int] = ListBuffer(0, 1, 2, 3)

scala> list.toArray
res5: Array[Int] = Array(0, 1, 2, 3)
```

3. 列表常用操作

Scala 也提供了针对列表的常用操作，比如，判断列表是否为空（isEmpty）。

```
scala> val list = List(0,1,2,3)
list: List[Int] = List(0, 1, 2, 3)

scala> list.isEmpty
res0: Boolean = false
```

示例：拼接两个列表（++），返回的是一个在内存中生成的新列表。

```
scala> val list = List(0,1,2,3)
list: List[Int] = List(0, 1, 2, 3)

scala> val tmpList = List(4,5)
tmpList: List[Int] = List(4, 5)

scala> list++tmpList
res1: List[Int] = List(0, 1, 2, 3, 4, 5)
```

示例：获取列表的首个元素（head）和剩余部分（tail）。head 用于获取首个元素，tail 用于获取除了第一个元素外的其余元素，返回的是一个在内存中生成的新列表。

```
scala> val list = List(0,1,2,3)
list: List[Int] = List(0, 1, 2, 3)

scala> list.head
res0: Int = 0

scala> list.tail
res1: List[Int] = List(1, 2, 3)
```

示例：反转（reverse）列表。

```
scala> val list = List(0,1,2,3)
list: List[Int] = List(0, 1, 2, 3)

scala> list.reverse
res2: List[Int] = List(3, 2, 1, 0)
```

示例：获取列表前缀（take）和后缀（drop）。

```
scala> val list = List(0,1,2,3)
list: List[Int] = List(0, 1, 2, 3)

scala> list.take(3)
res6: List[Int] = List(0, 1, 2)

scala> list.drop(1)
res7: List[Int] = List(1, 2, 3)
```

示例：扁平化（flatten）。扁平化适用于一个列表中包含多个列表的场景。通过扁平化可以将某个列表中包含的列表元素提取出来放在另一个列表中，如果列表中有相同的元素，会保留。

```
scala> val list = List(List(0),List(2,3),List(1),List(0))
list: List[List[Int]] = List(List(0), List(2, 3), List(1), List(0))

scala> list.flatten
res12: List[Int] = List(0, 2, 3, 1, 0)
```

示例：拉链（zip）。将两个列表通过相同的索引组成一个元素为元组的列表。比如，对一个学生姓名列表和班级列表做拉链，可以求得每个同学分配到的班级。

```
scala> val studentName = List("student1","student2","student3")
studentName: List[String] = List(student1, student2, student3)

scala> val studentClass = List("class1","class2","class3")
studentClass: List[String] = List(class1, class2, class3)

scala> studentName.zip(studentClass)
res13: List[(String, String)] = List((student1,class1), (student2,class2), (student3,class3))
```

示例：拉开（unzip）。将一个拉链的列表拉开，生成包含两个列表的元组。

```
scala> val zipList = studentName.zip(studentClass)
zipList: List[(String, String)] = List((student1,class1), (student2,class2), (student3,class3))
scala> zipList.unzip
res14: (List[String], List[String]) = (List(student1, student2, student3),List(class1, class2, class3))
```

示例：转换字符串（toString）。将一个列表转换成字符串。

```
scala> val list = List(0,1,2,3)
list: List[Int] = List(0, 1, 2, 3)

scala> list.toString
res0: String = List(0, 1, 2, 3)
```

示例：生成字符串（mkString）。将列表通过分隔符转换成字符串。

```
scala> val list = List(0,1,2,3)
list: List[Int] = List(0, 1, 2, 3)

scala> list.mkString("\t")
res2: String = 0	1	2	3
```

示例：并集（union）。将两个列表取并集，不去重。如果需要去重，要调用distinct。

```
scala> val list1= List(0,1,2,3)
list1: List[Int] = List(0, 1, 2, 3)

scala> val list2 = List(2,4,5,6)
list2: List[Int] = List(2, 4, 5, 6)

scala> list1.union(list2)
res3: List[Int] = List(0, 1, 2, 3, 2, 4, 5, 6)

scala> list1.union(list2).distinct    // 去重
res4: List[Int] = List(0, 1, 2, 3, 4, 5, 6)
```

示例：交集（intersect）。将两个列表取交集。

```
scala> val list1= List(0,1,2,3)
list1: List[Int] = List(0, 1, 2, 3)

scala> val list2 = List(2,4,5,6,3)
list2: List[Int] = List(2, 4, 5, 6, 3)
```

```
scala> list1.intersect(list2)
res5: List[Int] = List(2, 3)
```

示例：差集（diff）。list1.diff(list2) 表示获取 list1 中 list2 中不包含的元素。

```
scala> val list1= List(0,1,2,3)
list1: List[Int] = List(0, 1, 2, 3)

scala> val list2 = List(2,4,5,6,3)
list2: List[Int] = List(2, 4, 5, 6, 3)

scala> list1.diff(list2)
res6: List[Int] = List(0, 1)
```

1.8.3 Set 集合

Scala 和 Java 一样，也提供了没有重复元素的集合。Set 集合没有重复的元素，不保证插入的顺序。

同样，Set 集合也分为可变和不可变两种。定义可变 Set 集合前，要手动导入 Set 所在包的路径：import scala.collection.mutable.Set。

1. 不可变 Set 集合

创建不可变 Set 集合的语法格式如下。

```
// 创建一个空的不可变Set集合
val/var 变量名 = Set[类型]()
// 给定元素来创建一个不可变Set集合
val/var 变量名 = Set[类型](元素1,元素2,元素3,...)
```

示例：创建一个不可变 Set 集合。

```
scala> val set = Set(1,2,3,1,2,3)
set: scala.collection.immutable.Set[Int] = Set(1, 2, 3)
```

2. 可变 Set 集合

示例：定义一个可变 Set 集合，并添加新的元素到集合。

```
scala> val set = Set(1,2,3)
set: scala.collection.mutable.Set[Int] = Set(1, 2, 3)
```

```
scala> set += 4
res14: set.type = Set(1, 2, 3, 4)
```

示例：定义一个可变 Set 集合，并移除其中的一个元素。

```
scala> import scala.collection.mutable.Set
import scala.collection.mutable.Set

scala> val set = Set(1,2,3)
set: scala.collection.mutable.Set[Int] = Set(1, 2, 3)

scala> set -= 3
res16: set.type = Set(1, 2)
```

3. Set 集合常用操作

示例：获取 Set 集合大小（size）。

```
scala> val set = Set(1,2,3)
set: scala.collection.immutable.Set[Int] = Set(1, 2, 3)

scala> set.size
res7: Int = 3
```

示例：遍历 Set 集合（与数组一致）。

```
scala> val set = Set(1,2,3)
set: scala.collection.immutable.Set[Int] = Set(1, 2, 3)

scala> for(i <- set){
     |     println(i)
     | }
1
2
3
```

示例：添加一个元素，生成一个新的 Set 集合。

```
scala> val set = Set(1,2,3)
set: scala.collection.immutable.Set[Int] = Set(1, 2, 3)

scala> set + 4
res11: scala.collection.immutable.Set[Int] = Set(1, 2, 3, 4)
```

示例：拼接一个 Set 集合，生成一个新的 Set 集合。

```
scala> val set = Set(1,2,3)
set: scala.collection.immutable.Set[Int] = Set(1, 2, 3)

scala> set ++ Set(4,5)
res12: scala.collection.immutable.Set[Int] = Set(5, 1, 2, 3, 4)
```

示例：拼接一个列表，生成一个新的 Set 集合。

```
scala> val set = Set(1,2,3)
set: scala.collection.immutable.Set[Int] = Set(1, 2, 3)

scala> set ++ List(4,5)
res13: scala.collection.immutable.Set[Int] = Set(5, 1, 2, 3, 4)
```

1.9 Map 映射

和 Java 一样，键值对（key-value）的集合用 Map 表示，同样地，也分不可变 Map 和可变 Map。在定义可变 Map 前也要先手动导入 import scala.collection.mutable.Map。

1.9.1 不可变 Map

定义不可变 Map 的语法格式如下。

```
val/var map = Map(键->值，键->值，键->值,...)  // 推荐，可读性更好
val/var map = Map((键，值)，(键，值)，(键，值)，(键，值),...)
```

示例：定义一个学生成绩 Map 集合，其包含学生姓名和学生成绩，获取其中一个学生的成绩。

```
scala> val map = Map("student1"->99 , "student2"->98)
map: scala.collection.immutable.Map[String,Int] = Map(student1 -> 99, student2 -> 98)

scala> map("student1")
res17: Int = 99
```

1.9.2 可变 Map

定义之前手动导入 import scala.collection.mutable.Map，定义语法格式与不可变 Map 的一致。

示例：定义一个学生成绩 Map 集合，修改第一个学生的成绩为 100。

```
scala> import scala.collection.mutable.Map
import scala.collection.mutable.Map

scala> val map = Map("student1"-> 99 ,"student2"->98)
map: scala.collection.mutable.Map[String,Int] = Map(student2 -> 98, student1 -> 99)

scala> map("student1") = 100

scala> map
res19: scala.collection.mutable.Map[String,Int] = Map(student2 -> 98, student1 -> 100)
```

1.9.3 Map 基本操作

1. 获取值（map(key)）

此类方式适用于明确知道 key（键）是什么的场景。当不确定 key 是什么的时候，为避免报错，使用 getOrElse(key, defaultValue)，当不存在 key 的时候使用默认值。

示例：定义一个学生成绩 Map 集合，获取第一个学生的成绩。

```
scala> val map = Map("student1"-> 99 ,"student2"->98)
map: scala.collection.immutable.Map[String,Int] = Map(student1 -> 99, student2 -> 98)

scala> map("student1")
res0: Int = 99
```

示例：使用 getOrElse 获取值。

```
scala> val map = Map("student1"-> 99 ,"student2"->98)
map: scala.collection.immutable.Map[String,Int] = Map(student1 -> 99, student2 -> 98)

scala> map.getOrElse("student4","defaultValue")
res6: Any = defaultValue
```

示例：定义一个学生成绩 Map 集合，获取所有学生姓名。

```
scala> val map = Map("student1"-> 99 ,"student2"->98)
map: scala.collection.immutable.Map[String,Int] = Map(student1 -> 99, student2 -> 98)
scala> map.keys
res1: Iterable[String] = Set(student1, student2)
```

示例：定义一个学生成绩 Map 集合，获取所有学生成绩。

```
scala> val map = Map("student1"-> 99 ,"student2"->98)
map: scala.collection.immutable.Map[String,Int] = Map(student1 -> 99, student2 -> 98)
scala> map.values
res0: Iterable[Int] = MapLike(99, 98)
```

示例：定义一个学生成绩 Map 集合，打印所有学生和成绩。

```
scala> val map = Map("student1"-> 99 ,"student2"->98)
map: scala.collection.immutable.Map[String,Int] = Map(student1 -> 99, student2 -> 98)
scala> for((studentName,studentScore)<- map){
     |     println("studentName is " + studentName + ",studentScore is " + studentScore)
     | }
studentName is student1,studentScore is 99
studentName is student2,studentScore is 98
```

2. 增加键值对

示例：

```
scala> val map = Map("student1"-> 99 ,"student2"->98)
map: scala.collection.immutable.Map[String,Int] = Map(student1 -> 99, student2 -> 98)
```

```
scala> map + ("student3"->97)
res3: scala.collection.immutable.Map[String,Int] = Map(student1 -> 99, student2 -> 98, student3 -> 97)
```

3. 删除键值对

示例：

```
scala> val map = Map("student1"-> 99 ,"student2"->98)
map: scala.collection.immutable.Map[String,Int] = Map(student1 -> 99, student2 -> 98)
```

```
scala> map - "student2"
res4: scala.collection.immutable.Map[String,Int] = Map(student1 -> 99)
```

1.10 函数式编程

函数式编程是一种编程范式，其有自己的语言特性，如数据不可变、函数是第一公民等。

1.10.1 遍历（foreach）

对于数组，List、Set 等函数使用的遍历方式涉及 for 表达式，使用 foreach 函数式编程进行遍历，可以使程序更简洁。

```
foreach(f: (A) ⇒ Unit): Unit
```

参数：接收一个函数对象，函数的输入参数为集合的元素。

返回值: null。

foreach 执行过程：集合中的每一个元素都经过函数 f。

示例：定义一个列表，用 foreach 遍历。

```
scala> val list = List(1,2,3,4)
list: List[Int] = List(1, 2, 3, 4)
```

```
scala> list.foreach((x:Int)=>println(x))
```

1.10.2 使用类型推断简化函数定义

上面的示例可以用更简单的方式实现,利用类型推断简化函数定义,Scala 可以自动推断出集合中每个元素参数的类型,所以定义函数时,可以省略其参数列表的类型。

```
scala> val list = List(1,2,3,4)
list: List[Int] = List(1, 2, 3, 4)

scala> list.foreach(x=>println(x))
```

1.10.3 使用下画线简化函数定义

当函数参数只在函数体中出现一次,而且函数体没有嵌套调用时,可以使用下画线来简化函数的定义。将上面使用自动类型推断的示例继续简化,使用下画线代替函数,可以使打印语句更简练。

```
scala> val list = List(1,2,3,4)
list: List[Int] = List(1, 2, 3, 4)

scala> list.foreach(println(_))
1
2
3
```

1.10.4 映射(map)

如果想要集合中的每一个元素都经过同一个函数进行处理,可以使用 map。

```
def map[B](f: (A) ⇒ B): TraversableOnce[B]
```

泛型 B 指定最终返回的集合泛型;f:(A) ⇒ B 传入一个函数对象,该函数接收一个类型 A(要转换的列表元素),返回值为类型 B;TraversableOnce[B] 是 B 类型的集合。

示例:定义一个列表,并使其中每一个值都加 1。

```
scala> val list = List(0,1,2,3)
list: List[Int] = List(0, 1, 2, 3)

scala> list.map(x=>x+1)
res3: List[Int] = List(1, 2, 3, 4)
```

同样，使用下画线简化函数定义。

```
scala> val list = List(0,1,2,3)
list: List[Int] = List(0, 1, 2, 3)
scala> list.map(_+1)
res4: List[Int] = List(1, 2, 3, 4)
```

1.10.5 扁平化映射（flatMap）

可以把 flatMap 理解为先进行 map 操作，再进行 flatten 操作，即将列表中的元素转换成 List 进行处理后再"拍平"。

```
def flatMap[B](f: (A) ⇒ GenTraversableOnce[B]): TraversableOnce[B]
```

泛型 B 指定最终返回的集合泛型；f:(A) ⇒ B 传入一个函数对象，该函数接收一个类型 A（要转换的列表元素），返回值为类型 B；TraversableOnce[B] 是 B 类型的集合。

比如，一个列表中有多个文本行数据，每一元素均为一个班级的学生姓名，现在想要统计这个列表中姓名不一样的学生一共有多少，可使用 map+flatten 实现。

```
scala> val list = List("student1 student2 student3","student1 student4 student5")
list: List[String] = List(student1 student2 student3, student1 student4 student5)

scala> list.map(_.split(" ")).flatten.distinct
res8: List[String] = List(student1, student2, student3, student4, student5)

scala> list.map(_.split(" ")).flatten.distinct.length
res9: Int = 5
```

示例：使用 flatMap 实现。

```
scala> val list = List("student1 student2 student3","student1 student4 student5")
list: List[String] = List(student1 student2 student3, student1 student4 student5)
```

```
scala> list.flatMap(_.split(" "))
res13: List[String] = List(student1, student2, student3, student1, student4, student5)
```

```
scala> list.flatMap(_.split(" ")).distinct.length
res14: Int = 5
```

1.10.6 过滤（filter）

过滤符合一定条件的元素，并返回元素列表。

```
def filter(f: (A) ⇒ Boolean): TraversableOnce[A]
```

f:(A) ⇒ Boolean 传入一个函数对象，该函数接收一个类型 A（要转换的列表元素），返回一个 Boolean 类型，满足条件为 true，不满足为 false；TraversableOnce[A] 返回过滤条件等于 true 的元素，A 为元素类型。

示例：定义一个列表，其包含某班级的所有学生成绩，过滤成绩大于 80 的元素。

```
scala> val list = List(80,50,98,99,33,67,93)
list: List[Int] = List(80, 50, 98, 99, 33, 67, 93)
```

```
scala> list.filter(_>80)
res15: List[Int] = List(98, 99, 93)
```

1.10.7 排序

Scala 中提供了 3 种排序方式：默认排序、指定字段排序、自定义排序。

1. 默认排序（sorted）

示例：定义一个列表，其包含 1、2、3、4、5 元素，对其进行升序排列。

```
scala> val list = List(4,5,2,1,3)
list: List[Int] = List(4, 5, 2, 1, 3)
```

```
scala> list.sorted
res16: List[Int] = List(1, 2, 3, 4, 5)
```

2. 指定字段排序（sortBy）

可以指定按特定字段排序，将传入的函数转换后再进行排序。

```
def sortBy[B](f: (A) ⇒ B): List[A]
```

f:(A) ⇒ B 传入一个函数对象，该函数接收一个类型 A（要转换的列表元素），返回类型 B 的元素进行排序；List[A] 是返回的排序后的列表。

示例：定义一个列表，其包含 "student1 90"、"student2 87"、"student3 99"，并按学生成绩排序。

```
scala> val list =List("student1 90","student2 87","student3 99")
list: List[String] = List(student1 90, student2 87, student3 99)

scala> list.sortBy(_.split(" ")(1))
res18: List[String] = List(student2 87, student1 90, student3 99)
```

3. 自定义排序（sortWith）

自定义排序，是根据一个函数来进行自定义排序。

```
def sortWith(lt: (A, A) ⇒ Boolean): List[A]
```

lt:(A,A) ⇒ Boolean 传入一个比较大小的函数对象，该函数接收两个集合类型的元素参数，返回两个元素比较大小的结果，小于则返回 true，大于则返回 false。

示例：定义一个列表，其包含1、2、3、4、5元素，使用sortWith对其进行升序排列。

```
scala> val list = List(4,3,1,2,5)
list: List[Int] = List(4, 3, 1, 2, 5)

scala> list.sortWith((x,y) => if(x<y)true else false)
res19: List[Int] = List(1, 2, 3, 4, 5)
```

可以使用下画线来简化代码。

```
scala> val list = List(4,3,1,2,5)
list: List[Int] = List(4, 3, 1, 2, 5)
scala> list.sortWith(_ < _)
res20: List[Int] = List(1, 2, 3, 4, 5)      // 简化
```

1.10.8 分组（groupBy）

如果需要将数据分组后进行统计，那么使用 groupBy。

```
def groupBy[K](f: (A) ⇒ K): Map[K, List[A]]
```

[K] 是分组字段的类型；f: (A) ⇒ K 传入一个函数对象，该函数接收集合类型的元素参数，返回一个 K 类型的 key，这个 key 会用来进行分组，相同的 key 放在一组；Map[K, List[A]] 返回一个映射，K 为分组字段，List 为这个分组字段对应的一组数据。

示例：定义一个列表，其包含 "student1 90"、"student2 80"、"student3 80"，并按成绩进行分组。

```
scala> val list = List("student1 90","student2 80","student3 80")
list: List[String] = List(student1 90, student2 80, student3 80)

scala> list.groupBy(_.split(" ")(1))
res21: scala.collection.immutable.Map[String,List[String]] = Map(90 -> List(student1 90), 80 -> List(student2 80, student3 80))
```

1.10.9 聚合（reduce）

聚合操作可以将一个列表中的数据合并为一个，其在统计分析中很常用。

```
def reduce[A1 >: A](op: (A1, A1) ⇒ A1): A1
```

[A1 >: A] 中（下界）A1 必须是集合类型的子类；op: (A1, A1) ⇒ A1 传入一个函数对象，用来不断进行聚合操作，第一个 A1 类型参数为当前聚合后的变量，第二个 A1 类型参数为当前要进行聚合的元素；返回值 A1 是列表最终聚合成的一个元素。简单来说就是前一个元素和后一个元素聚合后，再与下一个元素聚合。

示例：定义一个列表，其包含 1、2、3、4、5、6、7、8、9、10，求和。

```
scala> val list = List(1,2,3,4,5,6,7,8,9,10)
list: List[Int] = List(1, 2, 3, 4, 5, 6, 7, 8, 9, 10)

scala> list.reduce(_+_)
res22: Int = 55    // 解析执行流程：1+2 =3 , 3+3 = 6, ...
```

可以指定其计算方向，默认是从左到右。

```
scala> val list = List(1,2,3,4,5,6,7,8,9,10)
list: List[Int] = List(1, 2, 3, 4, 5, 6, 7, 8, 9, 10)

scala> list.reduce(_+_)
res22: Int = 55

scala> list.reduceLeft(_+_)    // 从左到右：1+2 =3 , 3+3 = 6, ...
res23: Int = 55

scala> list.reduceRight(_+_)   // 从右到左：10+9 =19 , 19+8 = 27, ...
res24: Int = 55
```

1.10.10 折叠（fold）

fold 与 reduce 类似，但是多了一个初始值参数，在执行时会先指定初始值。

```
def fold[A1 >: A](z: A1)(op: (A1, A1) ⇒ A1): A1
```

泛型 [A1 >: A] 中（下界）A1 必须是集合类型的子类；z: A1 是初始值；op: (A1, A1) ⇒ A1 传入一个函数对象，用来不断进行折叠操作，第一个 A1 类型参数为当前折叠后的变量，第二个 A1 类型参数为当前要进行折叠的元素；返回值 A1 是列表最终折叠成的一个元素。

示例：定义一个列表，其包含 1、2、3、4、5、6、7、8、9、10，求和。

```
scala> val list = List(1,2,3,4,5,6,7,8,9,10)
list: List[Int] = List(1, 2, 3, 4, 5, 6, 7, 8, 9, 10)

scala> list.fold(0)(_+_)    // 从左到右：0+1=1 , 1+2 =3 , 3+3 = 6, ...
res25: Int = 55
```

可以指定其计算方向，默认是从左到右。

```
scala> val list = List(1,2,3,4,5,6,7,8,9,10)
list: List[Int] = List(1, 2, 3, 4, 5, 6, 7, 8, 9, 10)

scala> list.fold(0)(_+_)
res25: Int = 55

scala> list.foldLeft(0)(_+_)    // 从左到右：0+1=1 , 1+2 =3 , 3+3 = 6, ...
```

```
res26: Int = 55

scala> list.foldRight(0)(_+_)    // 从右到左：0+10 = 10 ,10+9 =19 , 19+8 = 27, ...
res27: Int = 55
```

1.11 本章总结

本章重点讲解了 Scala 的语法基础，其中重点为数据类型和集合类型。Scala 对比 Java 来说，不可变的特性贯穿了整个语法基础知识体系，包括定义变量时的 val 类型，默认定义集合为不可变型。想要定义可变集合，必须先手动引入 import scala.collection.mutable。且作为一个面向函数编程的语言，Scala 在大数据开发运用中，foreach、map、flatMap、reduce、groupBy、filter 等操作随处可见，读者要深入了解并熟练掌握。

1.12 本章习题

1. 完成 Scala 开发环境搭建，并掌握 Scala 语言的基本语法。
2. 完成数组、元组、映射和函数式编程的代码实践。

第 2 章 Scala 面向对象编程

Scala 是一门支持面向对象编程的语言，因此也有类与对象的概念。所以，我们可以基于 Scala 面向对象的特性来开发面向对象的应用程序。当然，我们可以通过 Scala 面向对象的 API（Application Programming Interface，应用程序接口）来调用 Java 语言中面向对象封装好的类的对象的动态方法和静态方法。

2.1 类与对象

Scala 和 Java 一样，可以通过 class 来创建类，而使用 class 创建的类需要通过 new 创建对象。

想要运行代码需要一个 main 方法，main 方法是程序的启动入口。需要在静态对象中创建 main 方法，Scala 中没有静态方法或者静态属性，使用单例对象 object 定义的方法或属性相当于使用 static 关键字定义的静态方法和属性。

比如，创建一个学生信息类，并使用该类创建学生对象。

```
// 创建一个学生信息类
class StudentInfo {}

// 想要运行代码，需要一个 main 方法
object Run {
```

```
  def main(args: Array[String]): Unit = {
    print(new StudentInfo())
  }
}
```

运行结果为 Part2.StudentInfo@146ba0ac，打印了这个对象的地址。这样就简单创建了一个对象。

在创建对象的时候，如果类是空的，没有任何成员变量，可以省略 {}。而构造参数为空的时候，可以省略 ()。那么上述创建对象的代码可以简写成如下形式。

```
class StudentInfo   // 简写：类为空，省略{}
//  想要运行代码，需要一个main 方法
object Run {
  def main(args: Array[String]): Unit = {
    print(new StudentInfo)   // 简写：构造参数为空，省略()
  }
}
```

同样打印对象地址 Part2.StudentInfo@146ba0ac。

2.2 定义和访问成员变量

在创建类时可以为这个类定义变量，在定义时通过 val 和 var 来控制变量是否可变。通过 val 定义的变量是不可变的，其没有提供 set 方法；通过 var 定义的变量是可变的，其提供了 set 方法。而 get 方法两种创建形式都会提供。

要获取属性值，直接调用属性名即可。对于使用 var 定义的属性，想要重新赋值，也可以直接调用属性名。

示例：创建一个学生信息类，其包含学生姓名、学生年龄、学生学号和班级信息，属性使用 val 和 var 来定义，尝试打印属性值，并尝试重新赋值。

```
class StudentInfo{
  val studentName:String = "student1"
  val studentAge:Int = 19
  val studentNum:Int = 1001
  val classInfo:String = "class1"
}
```

```
object Run {
  def main(args: Array[String]): Unit = {
    val studentinfo = new StudentInfo
    println(s"init data : studentName is ${studentinfo.studentName} ; studentAge is ${studentinfo.studentAge} ;" +
      s" studentNum is ${studentinfo.studentNum} ; student classInfo is ${studentinfo.classInfo}")
    studentinfo.studentName = "student2"   // 值
    studentinfo.studentAge = 20
    studentinfo.studentNum = 1002
    studentinfo.classInfo = "class2"
    println(s"update data : studentName is ${studentinfo.studentName} ; studentAge is ${studentinfo.studentAge} ;" +
      s" studentNum is ${studentinfo.studentNum} ; student classInfo is ${studentinfo.classInfo}")
  }
}
```

在运行的过程中，我们会看到报错，说明不能对 val 修饰的属性赋值。那么将上面代码中的 val 改成 var。

```
class StudentInfo{
  var studentName:String = "student1"
  var studentAge:Int = 19
  var studentNum:Int = 1001
  var classInfo:String = "class1"
}
```

运行结果如图 2-1 所示。

```
init data : studentName is student1 ; studentAge is 19 ; studentNum is 1001 ; student classInfo is class1
update data : studentName is student2 ; studentAge is 20 ; studentNum is 1002 ; student classInfo is class2
```

图 2-1

可以看出属性经过修改，变为了最新数据。

在定义的时候，可以应用 Scala 的特性，让 Scala 自动推断补齐类型。上述代码可以精简成如下代码。

```
class StudentInfo{
  var studentName = "student1"
  var studentAge = 19
```

```
    var studentNum = 1001
    var classInfo = "class1"
}
```

细心的朋友可能发现了,为什么我们定义属性的时候要初始化呢?为什么在 Java 中就可以不初始化呢?其实,在 Java 中定义属性时可以不初始化,这时值为默认值,而在 Scala 中是一定要初始化的,使用 val 定义的属性必须进行赋值,使用 var 定义的属性可以使用默认的初始化值。下面介绍默认初始化的方式。

2.3 使用下画线初始化成员变量

对于使用 var 定义的属性,在初始化时可以使用下画线来初始化默认值。不同类型数据的初始化其初始值不一样,在使用下画线初始化时,其类型不可省略。

示例:创建一个学生信息类,其包含学生姓名、学生年龄、学生学号、班级信息,使用 var 定义属性值为默认值。

```
class StudentInfo{
  var studentName = _      // 没有指定类型
  var studentAge = _       // 没有指定类型
  var studentNum = _       // 没有指定类型
  var classInfo = _        // 没有指定类型
}
```

上面没有指定类型是无法定义属性的,会报图 2-2 所示的错误。

```
unbound placeholder parameter
  var classInfo = _
```

图 2-2

为上述属性指定类型,并初始化数据。

```
class StudentInfo{
  var studentName:String = _    // 指定类型
  var studentAge:Int = _        // 指定类型
  var studentNum:Int = _        // 指定类型
  var classInfo:String = _      // 指定类型
}
```

```
object Run {
  def main(args: Array[String]): Unit = {
    val studentinfo = new StudentInfo
    println(s"init data : studentName is ${studentinfo.studentName} ; studentAge is ${studentinfo.studentAge} ;" +
      s" studentNum is ${studentinfo.studentNum} ; student classInfo is ${studentinfo.classInfo}")
  }
}
```

运行结果如图 2-3 所示。

```
init data : studentName is null ; studentAge is 0 ; studentNum is 0 ; student classInfo is null
```

图 2-3

2.4 定义成员方法

Scala 中创建的类不仅可以定义属性，也可以定义方法。使用 def 定义方法即可。

示例：创建一个学生信息类，其包含学生姓名、学生年龄、学生学号、班级信息，并定义一个查找学生姓名的方法，参数为学生学号，返回值为学生姓名。

```
class StudentInfo {
  var studentName: String = _
  var studentAge: Int = _
  var studentNum: Int = _
  var classInfo: String = _

  /**
   * 在类中定义一个方法
   * @return message
   */
  def sayHello(message:String): String = {
    message
  }
}
object Run{
  def main(args: Array[String]): Unit = {
    val student = new StudentInfo
```

```
    println(student.sayHello("hello"))
  }
}
```

运行结果如图 2-4 所示。

```
hello
```

图 2-4

2.5 访问修饰符

Scala 也可以像 Java 一样，通过设置属性的修饰符来控制是否被访问。但是 Scala 中没有 public 修饰符，默认不加修饰符的属性和方法都为 public。

Scala 提供的修饰符为（private、package 和特殊的 this。private 修饰的类对本包可见，修饰的属性和方法只能在本类、本类的子类和伴生对象中可见。package 修饰的类、属性、方法均在指定的方法中可见。特殊的 this，去除了伴生对象的访问权限，严格意义上只能在本类和本类的子类中可见。

private 作用于类时，被修饰的类只能被同包下的子类继承，且同包继承的类也必须是 private 修饰的，这样才不会改变原始父类的可见性。

示例：创建一个被 private 修饰的类，并创建一个它的子类。

```
package code

private class PrivateClass {
  val item:String = "private item"
}

class SmallPrivateClass extends PrivateClass   // 未用private 修饰

object Run4{
  def main(args: Array[String]): Unit = {
    val s = new SmallPrivateClass
    println(s.item)
  }
}
```

以上代码没有使用 private 修饰子类，运行后会报错，如图 2-5 所示。

```
private class PrivateClass escapes its defining scope as part of type code.PrivateClass
class SmallPrivateClass extends PrivateClass
```

图 2-5

将子类用 private 修饰，再运行，可以看到它能获取到父类里的属性值。

```
package code

private class PrivateClass {
  val item:String = "private item"
}

private class SmallPrivateClass extends PrivateClass  // 使用 private 修饰

object Run4{
  def main(args: Array[String]): Unit = {
    val s = new SmallPrivateClass
    println(s.item)
  }
}
```

运行结果如图 2-6 所示。

```
private item
```

图 2-6

使用 private 修饰属性/方法表示该属性/方法只对本类、本类的子类和伴生对象可见，其他均不可见。

示例：创建一个类，使用 private 修饰一个属性和一个方法，并在伴生对象中输出这个属性和调用这个方法。

```
package code

class PrivateDemo {  private var studentName: String = "student1"
  var studentAge: Int = _
  var studentNum: Int = _
  var classInfo: String = _

  /**
   * 在类中定义一个方法
```

```
     * @param message message
     * @return message
     */
    private def sayHello(message:String): String = {
      message
    }
}
object PrivateDemo{
  def main(args: Array[String]): Unit = {
    val student = new PrivateDemo
    println(student.sayHello("hello"))
    println(student.studentName)
  }
}
```

上述 sayHello 方法和 studentName 在伴生对象中是可以正常访问的，当我们换一个类去访问的时候，就会发现其是不可见的，如图 2-7 所示。

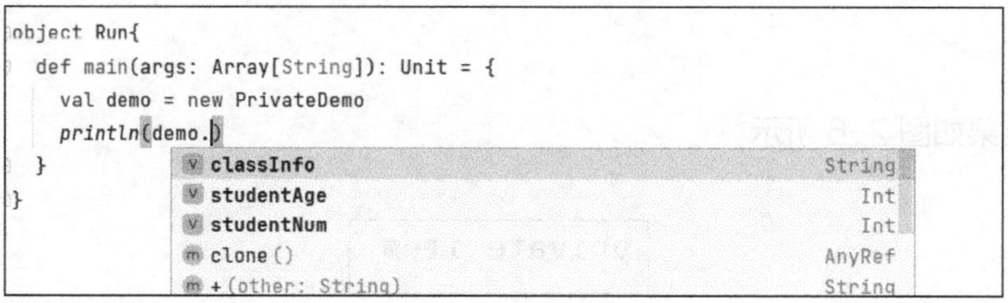

图 2-7

添加 this 修饰符后，会去除伴生对象的访问权限。

示例：创建一个类，使用 private 修饰一个属性和一个方法，并在其后加上 this，在伴生对象中输出这个属性和调用这个方法。

```
package code

class ThisDemo {
  private[this] var studentName: String = "student1"
  var studentAge: Int = _
  var studentNum: Int = _
  var classInfo: String = _

  /**
    * 在类中定义一个方法
```

```
 * @return message
 */
private[this] def sayHello(message:String): String = {
  message
}
}
```

当我们在伴生对象中调用其属性和方法时会发现找不到该属性和方法了，如图 2-8 所示。这就说明 this 修饰的属性和方法对伴生对象不可见。

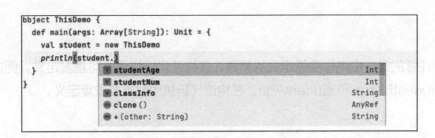

图 2-8

2.6 类的构造器

在创建类对象的时候，需要调用类的构造器。Scala 提供了默认的构造器，我们直接使用 new 创建即可。Scala 也提供了自定义的构造器，以便我们根据需求自定义创建。Scala 的自定义构造器分为主构造器和辅助构造器。

2.6.1 主构造器

Scala 提供的默认构造器没有参数，而主构造器有参数。

class 类名 {// 构造代码块 }

那么主构造器就是带参数的构造器。

class 类名 (var/val 参数名:参数类型 = 默认值, var/val 参数名:参数类型 = 默认值) {
**   // 构造代码块**
}

示例：创建一个学生信息类，其包含学生姓名、学生年龄、学生学号、班级信息，学生姓名和学生年龄由主构造器定义。

```scala
class StudentInfo(var studentName:String,var studentAge:Int) {
  var studentNum: Int = _
  var classInfo: String = _
}
object StudentInfo{
  def main(args: Array[String]): Unit = {
    val studentInfo = new StudentInfo("student1",20)
    System.out.print(studentInfo.studentName+": age is " + studentInfo.studentAge)
    System.out.print(studentInfo.studentNum+": age is " + studentInfo.classInfo)

  }
}
```

需要注意的是，由主构造器定义的参数，在构造代码块里不可重复定义。例如，上面代码的 studentName 和 studentAge，在构造代码块里就不能重复定义。

2.6.2 辅助构造器

如果需要使用多种方式创建对象，可以使用辅助构造器。定义辅助构造器和定义方法类似，都通过 def 关键字来定义，特殊的是，这个方法的名称是 this。

```scala
def this(参数名:参数类型，参数名:参数类型) {
// 第一行需要调用主构造器或者其他构造器
this() // 主构造器
// 构造代码块
}
```

需要注意，辅助构造器的第一行必须调用主构造器或者其他构造器。

示例：创建一个学生信息类，其包含学生姓名、学生年龄、学生学号、班级信息，学生姓名和学生年龄由辅助构造器定义。

```scala
class StudentInfo(){
  var studentName:String =_
  var studentAge:Int=_
  var studentNum: Int = _
  var classInfo: String = _

  def this(studentName:String,studentAge:Int){
    this()
    this.studentName = studentName
```

```
    this.studentAge = studentAge
  }
}

object StudentInfo{
  def main(args: Array[String]): Unit = {
    val studentInfo = new StudentInfo("student1",20)
    System.out.print(studentInfo.studentName+": age is " + studentInfo.studentAge)
    System.out.print(studentInfo.studentNum+": age is " + studentInfo.classInfo)
  }
}
```

由辅助构造器定义的参数，在构造代码块里可重复定义，这和主构造器有所区别。

2.7 单例对象

Scala 中没有 Java 中的静态成员，所以当我们想用 Scala 定义一个静态方法或静态属性的时候，就需要使用单例对象 object。单例对象像 Java 中的静态类，它的成员、方法都是静态的。

2.7.1 定义单例对象

单例对象表示全局有且仅有一个对象，类似于普通对象定义方式，将 class 变成 object。object 的属性和方法均可直接引用。定义的属性一般情况下均为大写。

示例：定义一个静态的学生信息单例对象，同样包含学生姓名、学生年龄、学生学号、班级信息，并打印学生信息。

```
object StudentInfo {
  var STUDENTNAME:String ="student1"
  var STUDENTAGE:Int= 19
  var STUDENTNUM: Int = 1001
  var CLASSINFO: String = "class1"

  def main(args: Array[String]): Unit = {
    System.out.println(StudentInfo.STUDENTNAME)
    System.out.println(StudentInfo.STUDENTAGE)
    System.out.println(StudentInfo.STUDENTNUM)
```

```
        System.out.println(StudentInfo.CLASSINFO)
    }
}
```

2.7.2 在单例对象中定义成员方法

在单例对象中定义的成员是静态的,可以直接访问;定义的方法也是静态的,同样可以直接访问。和定义普通的类相似,定义成员方法使用 def 关键字。

示例:定义一个静态的学生信息单例对象,并调用其静态方法 sayHello。

```
object StudentInfo {
  def sayHello(message:String): String = {
    message
  }
  def main(args: Array[String]): Unit = {
    StudentInfo.sayHello("hello")
  }
}
```

2.7.3 工具类案例

静态的方法和属性适合工具类,比如时间转换、连接数据库和字符串校验等。

示例:利用字符串校验工具类,检验字符串是否为空。

```
object StringUtil {
  /**
   * 判断传入的字符串是否为空
   * @param str  传入的字符串
   * @return
   */
  def isNotEmpty(str:String):Boolean = {
    str != null && !"".equals(str)
  }
  def main(args: Array[String]): Unit = {
    val str = "is not empty string"
    System.out.println(isNotEmpty(str))
  }
}
```

2.8　main 方法

所有程序的入口都是 main 方法，main 方法作为一个静态方法，必须放在 object 对象中。

2.8.1　定义 main 方法

定义 main 方法的语法格式如下：

```
def main(args: Array[String]): Unit = {
    // 代码块
}
```

示例：创建一个单例对象，并在单例对象中打印 hello world。

```
object Demo {
  def main(args: Array[String]): Unit = {
    System.out.println("hello world")
  }
}
```

2.8.2　实现 App trait 来定义入口

创建一个单例对象，继承使用 trait 关键字定义的 API，然后将需要编写在 main 方法中的代码写在单例对象的方法体内。

```
object 单例对象名 extends App {
// 方法体
}
```

示例：创建一个单例对象并实现 App trait（特质），在单例对象中打印 hello world。

```
object Demo extends App{
    System.out.println("hello world")

}
```

2.9 伴生对象

如果想让一个类使用静态方法或属性，又需要实例成员，可以使用伴生对象来实现。类似 Java，伴生对象不仅提供了便携的构建方法，同时也因为 Scala 运行在 JVM 上，与 Java 类库完全兼容的特点，静态属性是属于类的，在 JVM 中只开辟一块内存空间，很节省内存资源。并且由于只有一块内存空间，说明所有实例共享内存信息。

在外部访问对象时，直接使用伴生对象名.属性或方法即可，不需要使用 new。

2.9.1 定义伴生对象

如果在同一个 .class 文件中，一个类与该对象的名字一样，那么称该对象为该类的伴生对象，则该类为该对象的伴生类。这个类可以用该对象的名字访问到所有成员。伴生类和伴生对象可以互相访问彼此的 private 属性。

示例：定义一个学生信息类，并创建其伴生对象，打印学生信息。

```
class StudentInfo {
  /**
   * 在伴生类中定义一个方法
   * @param message message
   * @return message
   */
  private def sayHello(message:String): String = {
    message
  }
}

/**
 * 伴生对象
 */
object StudentInfo{
  // 伴生对象中的私有属性
  private val message = "hello world"
  def main(args: Array[String]): Unit = {
    val student = new StudentInfo
```

```
        println(student.sayHello(message))
    }
}
```

我们还可以使用伴生对象快速创建对象：

```
val arr= Array("1","2","3")
```

2.9.2 apply 和 unapply 方法

apply 和 unapply 属于 Scala 的语法糖，在 class 的伴生对象里面定义。apply 方法的主要作用是让我们像调用方法一样创建对象，而 unapply 方法主要和 match 一起使用。

1. apply 方法

通过伴生对象快速创建的对象 val arr= Array("1","2","3") 原理上是怎么实现的呢？为什么可以不用 new 创建对象呢？

```
object 伴生对象名 {
def apply(参数名:参数类型，参数名:参数类型,...) = new 类(...)
}
```

或

```
class 伴生对象名 {
def apply(参数名:参数类型，参数名:参数类型,...) = new 类(...)
}
```

两种 apply 方法的位置不同，实际上在创建对象和使用上也不同。

示例：通过实现 apply 方法创建学生信息类。

```
/**
 * 伴生类
 * @param studentName
 * @param studentAge
 */
class StudentInfo(val studentName:String,val studentAge:Int){
    def apply(studentName: String, studentAge: Int): StudentInfo = new StudentInfo(studentName, studentAge)
}
/**
```

```
 * 伴生对象
 */
object StudentInfo{
    // 使用apply方法
    def apply(studentName:String,studentAge:Int): StudentInfo = new StudentInfo(studentName,studentAge)
    // 伴生对象中的private属性
    private val message = "hello world"
    def main(args: Array[String]): Unit = {
        val studentInfo = StudentInfo("student1",19) // 调用的是 object 中的apply方法
        println(studentInfo.studentName)
        val student2 = studentInfo("student2",20) // 调用的是 class 中的apply方法
        println(student2.studentName)
    }
```

运行结果如图 2-9 所示。

```
student1
student2
```

图 2-9

2. unapply 方法

unapply 方法相当于 apply 方法的逆过程，对数据进行解析，主要和 match 一起使用。

示例：

```
package code

/**
 * 伴生类
 * @param studentName
 * @param studentAge
 */
class StudentInfo(val studentName:String,val studentAge:Int){
    def apply(studentName: String, studentAge: Int): StudentInfo = new StudentInfo(studentName, studentAge)

    def unapply(arg: StudentInfo): Option[(String, Int)] = {
        if ( studentName == null)
```

```
      None
    else Some(arg.studentName, arg.studentAge)
  }
}
/**
 * 伴生对象
 */
object StudentInfo{
  // 使用apply 方法
  def apply(studentName:String,studentAge:Int): StudentInfo = new StudentInfo(studentName,studentAge)
  // 伴生对象中的private属性
  private val message = "hello world"
  def main(args: Array[String]): Unit = {
    val studentInfo = StudentInfo("student1",19) // 调用的是 object 中的apply方法
    println(studentInfo.studentName)
    val student2 = studentInfo("student2",20) // 调用的是 class 中的apply方法
    println(student2.studentName)
    studentInfo match { // 调用的是 class 中的unapply方法
      case studentInfo(name, age) => println(name, age)
      case _ => None
    }
  }
}
```

运行结果如图 2-10 所示。

```
student1
student2
(student1,19)
```

图 2-10

2.10 继承

Scala 中的继承方式和 Java 中的继承方式一样，都使用 extends，目的是复用代码，提高开发效率。类和单例对象都可以继承父类。可以在子类中定义父类没有的属性和方法，可以重写父类方法。

2.10.1 定义语法

Scala 中定义继承的语法格式如下。

```
class/object 类名 extends 父类{
}
```

2.10.2 类继承

示例：定义一个学生信息类，并继承 User 类。

```
package code

class User{
  val username:String = "username"
  val age:Int = 19
  def getInfo(): String ={
    username + " is " + age + " old"
  }

}


/**
 * 继承User类
 */
class StudentInfo extends User

object Demo {

  def main(args: Array[String]): Unit = {
    val student = new StudentInfo
    // 调用父类方法
    println(student.getInfo())
  }
}
```

2.10.3 单例对象继承

示例：定义一个学生信息单例对象，并继承 User 类。

```scala
class User{
  val username:String = "username"
  val age:Int = 19
  def getInfo(): String ={
    username + " is " + age + " old"
  }

}
object StudentInfo extends User
object Demo {

  def main(args: Array[String]): Unit = {
    // 调用父类方法
    println(StudentInfo.getInfo())
  }
}
```

2.10.4 override 和 super

与 Java 类似，想要重写父类成员，需要使用 override，并使用 super 关键字来调用父类。

示例：存在一个 User 类，为父类，包含用户姓名和年龄两个字段，提供用于获取用户信息的方法。定义一个学生信息类继承该类，并重写用户姓名和年龄，重写用于获取用户信息的方法。

```scala
class User{
  val username:String = "username"
  val age:Int = 19
  def getInfo: String ={
    username + " is " + age + " old"
  }
}
```

```
/**
 * 重写父类方法
 */
class StudentInfo extends User {
  override val username:String = "studentname"
  override val age: Int = 19
  override def getInfo: String = "new info message is " + super.getInfo
}

object Demo {

  def main(args: Array[String]): Unit = {
    val student = new StudentInfo
    // 调用父类方法
    println(student.getInfo)
  }
}
```

2.11 类型判断

Scala 虽然提供了自动类型推断的功能，但是我们在开发时，有时需要根据类型进行判断，所以需要了解如何获取类型。Scala 提供了两个方法：isInstanceOf 和 getClass/classOf（见表 2-1）。

表 2-1 Scala 与 Java 对比

Scala	Java
isInstanceOf[Class]	obj instanceof Class
getClass	(Class)obj
classOf[Class]	Class.class

2.11.1 isInstanceOf 和 asInstanceOf 方法

Java 提供了 instanceof 关键字来判断一个对象是否是某一个类的实例，而 Scala 使用 isInstanceOf 和 asInstanceOf 方法。isInstanceOf 方法用于判断对象是否为指定类型，asInstanceOf 方法用于将对象转换为指定类型。

```
// 判断对象是否为指定类型
val trueOrFalse:Boolean = 对象.isInstanceOf[类型]
// 将对象转换为指定类型
val 变量 = 对象.asInstanceOf[类型]
```

示例：存在一个 User 类，定义一个 StudentInfo 类型，继承 User 类，创建 StudentInfo 对象，并指明它的类型是 User，判断该对象是否为 StudentInfo 类型，如果是，将其转换为 StudentInfo 类型并打印该对象。

```
class User
class StudentInfo extends User
object Demo {
  def main(args: Array[String]): Unit = {
    val student:User = new StudentInfo
    // 判断 student 是否为 StudentInfo 类型
    if(student.isInstanceOf[StudentInfo]) {
      // 将 student 转换为 StudentInfo 类型
      val student2 = student.asInstanceOf[StudentInfo]
      println(student2)
    }
  }
}
```

2.11.2 getClass 和 classOf

isInstanceOf 只能判断对象是否为指定类及其子类的对象，而不能精确判断对象就是指定类的对象。如果要精确地判断出对象就是指定类的对象，那么只能使用 getClass 和 classOf。

对象.getClass 可以精确获取对象的 Class 类对象，classOf[x] 可以精确获取 x 类的 Class 类对象，这样就可使用 == 运算符直接比较类型，具体操作见如下示例。

示例：存在一个 User 类，定义一个学生信息类，继承 User 类，创建一个对象，并指明它的类型是 User，使用 isInstanceOf 判断该对象是否为 User 类型，使用 getClass/classOf 判断该对象是否为 User 类型，使用 getClass/classOf 判断该对象是否为 StudentInfo 类型。

```
class User
class StudentInfo extends User
object Demo {
```

```
def main(args: Array[String]): Unit = {
  val student:User = new StudentInfo
  // 判断 student 是否为 User 类型
  println(student.isInstanceOf[User]) // true
  println(student.getClass == classOf[User]) //false
  println(student.getClass == classOf[StudentInfo])//true
}
```

2.12 抽象类

抽象类使用 abstract 关键字定义。抽象类包含抽象方法和非抽象方法。抽象类不允许多重继承，即一个类只能继承一个抽象类。如果变量没有初始化、方法没有方法体，那么我们称这个类为抽象类。

子类在继承的时候必须实现抽象成员，实现时可不使用 override。

定义抽象类

定义抽象类的语法格式如下。

```
abstract class 抽象类名 {
// 定义抽象字段
val 抽象字段名:类型
// 定义抽象方法
def 方法名(参数名:参数类型,参数名:参数类型,...):返回类型
}
```

示例：定义一个抽象类 User，学生信息类继承 User 类，打印 User 类中的重写方法。

```
package code

/**
 * 继承抽象类
 * @author Administrator
 */
abstract class User{
  var username:String
```

```
    val age:Int
    def sayHello:String
}

class StudentInfo extends User {
    //在Scala中覆盖一个抽象类中的字段时，可以不写override
    var username:String = "studentname"    //使用var定义的抽象字段只能使用var覆盖
    val age:Int = 18    //使用val定义的抽象字段只能使用val覆盖
    def sayHello:String = "hello world"    //使用def定义的方法可以使用val或者def覆盖
}

object Demo {
    def main(args: Array[String]): Unit = {
    /*
    1. 抽象类不一定有抽象字段，只需要添加abstract关键字；
    2. 有抽象字段的类一定是抽象类；
    3. 重写字段的本质是重写字段的setter、getter方法
    */
    val student = new StudentInfo
    println(student.sayHello)
    }
}
```

2.13 匿名内部类

匿名内部类是没有名称的子类，直接用来创建实例对象。Spark的源码中大量使用匿名内部类，其使用方法和Java中的一致。

```
val/var 变量名 = new 类/抽象类 {
// 重写方法
}
```

示例：创建一个User抽象类，并添加一个sayHello抽象方法，使用匿名内部类调用其方法。

```
abstract class User {
    def sayHello()
}
object Demo {
    def main(args: Array[String]): Unit = {
```

```
    // 直接用 new 来创建一个匿名内部类对象
    val user = new User {
      override def sayHello(): Unit = println("hello wold")
    }
    user.sayHello()
  }
}
```

2.14 特质

对标 Java 接口，Scala 提供了特质（trait）。特质是 Scala 中代码复用的基础单元，它可以将方法和字段定义封装起来，然后添加到类中。与类继承不一样的是，类继承要求每个类都只能继承一个超类，而一个类可以添加多个特质。与 Java 不同，这里不用 implement，在 Scala 中，无论继承类还是继承特质都是用 extends 关键字。

特质的定义和抽象类的定义很像，但它使用 trait 关键字。

```
trait 名称 {
  // 抽象字段
  // 抽象方法
}
```

继承方式如下：

```
class 类 extends 特质1 with 特质2 {
  // 字段实现
  // 方法实现
}
```

如果要继承多个特质，则使用 with 关键字。

2.14.1 trait 作为接口使用

trait（特质）作为接口使用的方法和 Java 中的一致。

示例：继承单个 trait。

```
trait Tyre{
  def run: Unit ={
```

```
    println("我是车轮")
  }
}
class Car extends Tyre{
  def display(): Unit ={
    println("一辆车")
  }
}
object Roadster extends App{
  var roadster = new Car
  roadster.display()
  roadster.run
}
```

示例：继承多个 trait。

```
package code



trait Tyre{
  def run: Unit ={
    println("我是车轮")
  }
}


trait Wheel{
  def control: Unit ={
    println("我是方向盘")
  }
}


//同时继承多个trait
class Car extends Tyre with Wheel{
  def display(): Unit ={
    println("这是一辆车")
  }
}


object Roadster extends App{
  var roadster = new Car
  roadster.display()
  roadster.run
```

```
    roadster.control
}
```

示例：使用对象继承 trait。

```
trait Tyre{
  def run: Unit ={
    println("我是车轮")
  }
}

trait Wheel{
  def control: Unit ={
    println("我是方向盘")
  }
}
object Car extends Tyre with Wheel{
  def display(): Unit ={
    println("这是一辆车")
  }
  def main(args: Array[String]): Unit = {
    Car.display()
    Car.run
    Car.control
  }
}
```

2.14.2　trait 中定义具体的字段和抽象字段

继承 trait 的子类自动拥有 trait 中定义的字段，字段直接被添加到子类中。

示例：通过 trait 来实现一个小汽车的购买时间工具类，该工具类可以自动添加小汽车的购买时间。

```
package code

import java.text.SimpleDateFormat
import java.util.Date



trait Tyre{
```

```scala
  val buyDate = new SimpleDateFormat("yyyy-MM-dd HH:mm")
  def run: Unit ={
    println("我是车轮")
  }
  def buyTime()
}


trait Wheel{
  def control: Unit ={
    println("我是方向盘")
  }
}

//同时继承多个trait
class Car extends Tyre with Wheel{
  def display(): Unit ={
    println("这是一辆车")
  }
  override def buyTime(): Unit = {
    val buyTime = s"${buyDate.format(new Date())}"
    println(buyTime)
  }
}

object Demo extends App{
  var car = new Car
  car.display()
  car.buyTime()
  car.run
  car.control
}
```

2.14.3 使用 trait 实现模板模式

在一个 trait 中，具体方法依赖于抽象方法，而抽象方法可以放到继承 trait 的子类中实现，这种设计方式称为模板模式。模板模式多用于工具类，实现不同需求。

在 Scala 中，trait 中可以定义抽象方法，也可以定义具体方法。抽象方法没有方法体，具体方法有方法体。没有方法体的方法在继承时必须要实现。

示例：实现 trait 中定义的抽象方法。

```scala
package code

import java.text.SimpleDateFormat
import java.util.Date


trait CarTmp{
  val buyDate = new SimpleDateFormat("yyyy-MM-dd HH:mm")
  def run(): Unit ={
    println("我有车轮")
  }
  def control(): Unit ={
    println("我有方向盘")
  }
  def buyTime()
}

//继承trait
class LitterCar extends CarTmp{
  override def buyTime(): Unit = {
    val buyTime = s"${buyDate.format(new Date())}"
    println(buyTime)
  }
}

object Demo extends App{
  var car = new LitterCar
  car.buyTime()
  car.run()
  car.control()
}
```

2.14.4 对象混入 trait

Scala 中可以将 trait 混入对象中，就是将 trait 中定义的方法、字段添加到对象中。

val/var 对象名 = new 类 with 特质

示例：单独为某个对象加 trait，添加额外属性等。

```scala
package code
import java.text.SimpleDateFormat
import java.util.Date
trait Price{
  def getPrice(price:Int): Unit ={
    println("这辆车的价格是: " + price)
  }
}

trait CarTmp{
  val buyDate = new SimpleDateFormat("yyyy-MM-dd HH:mm")
  def run(): Unit ={
    println("我有车轮")
  }
  def control(): Unit ={
    println("我有方向盘")
  }
  def buyTime()
}

//继承trait
class LitterCar extends CarTmp{
  override def buyTime(): Unit = {
    val buyTime = s"${buyDate.format(new Date())}"
    println(buyTime)
  }
}

object Demo extends App{
  var car = new LitterCar with Price
  car.getPrice(10000)
  car.buyTime()
  car.run()
  car.control()
}
```

2.14.5 使用 trait 实现调用链模式

类继承多个 trait 后，可以依次调用多个 trait 中的同一个方法，只要让多个 trait 中的同一个方法在最后都依次执行 super 关键字即可。在类中调用多个 trait 中的同一个方法时，首先会从最右边的 trait 中的方法开始执行，然后依次往左执行，形成一个调用链条。

只要中间有一个方法没有 super，往左的 trait 中的方法都不会再执行；当最左边的 trait 执行完方法后，才轮到 super 执行共同的方法。或者可以理解为，右边的 super 调用它左边的 trait 中的方法，直到最左边的完成。

2.14.6　trait 调用链

示例：模拟银行支付流程，定义一个交付链。

```
package code


trait BaseTrait {
  def handle(data:String) = println("支付完成...")
}
trait DataTrait extends BaseTrait {
  override def handle(data:String): Unit = {
    println("验证数据...")
    super.handle(data)
  }
}


trait SignatureTrait extends BaseTrait {
  override def handle(data: String): Unit = {
    println("校验签名...")
    super.handle(data)
  }
}


class PayService extends DataTrait with SignatureTrait {
  override def handle(data: String): Unit = {
    println("准备支付...")
    super.handle(data)
  }
}
object Demo{
  def main(args: Array[String]): Unit = {
    val service = new PayService
    service.handle("支付100万元")
```

}
}

运行结果如图 2-11 所示。

```
准备支付...
校验签名...
验证数据...
支付完成...
```

图 2-11

我们去掉校验签名调用的 super，观察一下结果是否有变化。

```
package code

trait BaseTrait {
  def handle(data:String) = println("支付完成...")
}
trait DataTrait extends BaseTrait {
  override def handle(data:String): Unit = {
    println("验证数据...")
    super.handle(data)
  }
}

trait SignatureTrait extends BaseTrait {
  override def handle(data: String): Unit = {
    println("校验签名...")
//    super.handle(data)
  }
}


class PayService extends DataTrait with SignatureTrait {
  override def handle(data: String): Unit = {
    println("准备支付...")
    super.handle(data)
  }
}
object Demo{
  def main(args: Array[String]): Unit = {
    val service = new PayService
```

```
    service.handle("支付100万元")
  }
}
```

运行结果如图 2-12 所示。

```
准备支付...
校验签名...
```

图 2-12

可以看到，后续继承的 trait 中的方法不会被调用。

2.14.7 trait 的构造机制

如果一个类实现了多个 trait，这些 trait 是如何构造的呢？ trait 也有构造代码，但和类不一样，trait 不能有构造器参数。每个 trait 只有一个无参数的构造器。一个类继承另一个类以及多个 trait，当创建该类的实例时，它的构造顺序如下。

① 执行父类的构造器。

② 从左到右依次执行 trait 的构造器。

③ 如果 trait 有父 trait，先构造父 trait；如果多个 trait 有同样的父 trait，则只初始化一次。

④ 执行子类的构造器。

2.14.8 trait 继承类

trait 也可以继承类，trait 会将类中的成员都继承下来。

示例：定义一个 trait，该 trait 继承一个类。

```
package code

class PriceTmp{
  def getPrice(price:Int): Unit ={
    println("price is " + price)
```

```
    }
}

trait WheelPrice extends PriceTmp {
  def control(wheelPrice:Int)=getPrice(wheelPrice)
}

class CarPrice extends WheelPrice{
  def getCar(wheelPrice:Int)= control(wheelPrice)
}
object Demo{
  def main(args: Array[String]): Unit = {
    val car = new CarPrice
    car.getCar(10000)
  }
}
```

运行结果如图 2-13 所示。

```
price is 10000
```

图 2-13

2.15 本章总结

本章主要介绍了 Scala 面向对象编程的基础知识，包含类与对象、定义和访问成员变量、定义成员方法、访问修饰符、类的构造器、单例对象、继承、抽象类以及匿名内部类等，通过示例代码展现面向对象编程的具体应用。

2.16 本章习题

1. 总结 Scala 面向对象编程的核心语法结构。

2. 加强面向对象编程代码实践。

3. 掌握 Scala 面向对象编程中的类与对象、伴生对象、匿名内部类等重难点。

第 3 章 Scala 编程高级应用

本章将主要介绍样例类、模式匹配、偏函数以及正则表达式的使用方法，重点分析非变、协变、逆变和上下界的相关概念。Actor 并发编程模型也是需要重点理解和掌握的内容。

3.1 样例类

样例类是一种特殊类，它可以用来快速定义一个用于保存不可变的数据的类（类似于 Java POJO 类），能够被优化以用于模式匹配，而且它会自动生成 apply 方法，允许我们快速地创建样例类实例对象。在并发编程和 Spark、Flink 框架中也会经常使用它。

3.1.1 定义样例类

定义样例类的语法格式如下。

```
case class 样例类名(成员变量名1:类型1, 成员变量名2:类型2，成员变量名3：类型3)
```

示例：定义一个样例类。

```
// 定义一个样例类
// 样例类有两个成员变量name、age
case class CaseUser(name:String, age:Int)
```

```
// 使用 var 指定成员变量是可变的
case class CaseStudent(var name:String, var age:Int)
object CaseClassDemo {
def main(args: Array[String]): Unit = {
// 1. 使用 new 创建实例
val user1 = new CaseUser("student1", 18)
println(user1)
// 2. 使用类名直接创建实例
val user2 = CaseUser("student2", 20)
println(user2)
// 3. 样例类默认的成员变量都是val类型, 除非手动指定变量为var类型
//user2.age = 22
// 编译错误! age默认为val类型
val student= CaseStudent("student3", 23)
student.age = 24
println(student)
```

3.1.2 样例类方法

1. toString 方法

toString 方法返回的格式如下。

```
case class CaseUser(name:String, age:Int)
object CaseClassDemo {
 def main(args: Array[String]): Unit = {
 val user= CaseUser("user", 21)
 println(user.toString)
 // 输出: CaseUser(user,21)
```

2. equals 方法

样例类自动实现了 equals 方法, 可以直接使用 == 比较两个样例类是否相等, 即所有的成员变量是否相等。

```
val user2= CaseUser("user2", 21)
val user3= CaseUser("user2", 21)
println(user2 == user3)
// 输出: true
```

3. hashCode 方法

样例类自动实现了 hashCode 方法，如果所有成员变量的值相同，则 hash 值相同，只要有一个成员变量的值不一样，则 hash 值不一样。

```
val user2= CaseUser("user2", 21)
val user3 = CaseUser("user2", 22)
println(user2.hashCode())
println(user3.hashCode())
```

4. copy 方法

样例类实现了 copy 方法，可以快速创建一个相同的实例对象，可以使用带名参数指定给成员进行重新赋值。

```
val user= CaseUser("user", 21)
val user2= lisi1.copy(name="user2")
println(user2)
```

3.1.3 样例对象

使用 case object 可以创建样例对象。样例对象是单例的，而且它没有主构造器。样例对象是可序列化的。

case object 样例对象名

样例对象主要用于定义枚举。

示例：定义枚举。

```
trait Sex /*定义一个Sex特质*/
case object Male extends Sex // 定义一个样例对象并继承Sex（也可用gender）特质
case object Female extends Sex
case class User(name:String, sex:Sex)
object CaseClassDemo {
    def main(args: Array[String]): Unit = {
        val user = User("张三", Male)
        println(user)
    }
}
```

示例：定义消息。

```
case class StartSendMessage(msg: String)
// 消息如果没有任何参数，就可以定义为样例对象
case object StopSendMessage
case object PauseSendMessage
case object ResumeSendMessage
```

样例类可以使用类名 (成员变量1, 成员变量2,...) 快速创建样例对象。定义样例类成员变量时，可以指定 var 类型，表示可变（默认是不可变的）。样例类自动生成了 toString、equals、hashCode、copy 方法。样例对象没有主构造器，可以使用样例对象来创建枚举或者标识一类没有任何数据的消息。

3.2 模式匹配

Scala 中有一个非常强大的模式匹配机制，可以应用在很多场景，如类 Java 语言中的 switch 语句、类型查询，以及快速获取数据。

3.2.1 简单匹配

Java 中有 switch 关键字，可以简化 if 条件判断语句。在 Scala 中，可以使用 match 表达式替代。

```
变量 match {
  case "常量1" => 表达式1
  case "常量2" => 表达式2
  case "常量3" => 表达式3
  case _ => 表达式4  // 默认匹配
}
```

下面是一个简单的示例。

```
println("请输出一个词: ")
// StdIn.readLine 表示从控制台读取一行文本
val line= StdIn.readLine()
val result = line match {
  case "one" => s"$line:number is 1"
  case "two" => s"$line:number is 2 "
```

```
    case "three" => s"$line:number is 4"
    case _ => s"未匹配到$line"
}
println(result)
```

match 表达式是有返回值的,可以将 match 表达式赋值给其他的变量。

3.2.2 守卫

在 Java 中只能简单地添加多个 case 语句,比如要匹配 0 ~ 7,就需要写出 8 个 case 语句。

```
int a = 0;
switch(a) {
  case 0: a += 1;
  case 1: a += 1;
  case 2: a += 1;
  case 3: a += 1;
  case 4: a += 2;
  case 5: a += 2;
  case 6: a += 2;
  case 7: a += 2;
  default: a = 0; }
```

在 Scala 中,可以使用守卫来简化上述代码,也就是在 case 语句中添加 if 条件判断语句。

```
println("请输入一个数字:")
var line = StdIn.readInt()
line match {
  case res if line >= 0 && line <= 3 => a += 1
  case res2 if line > 3 && line < 8 => a += 2
  case _ => line = 0
}
println(line)
```

3.2.3 匹配类型

match 表达式还可以进行类型匹配。

```
变量 match {
  case 类型1 变量名: 类型1 => 表达式1
```

```
case 类型2 变量名: 类型2 => 表达式2
case 类型3 变量名: 类型3 => 表达式3
...
case _ => 表达式4 }
```

下面是一个简单的示例。

```
// stripMargin 表示删除前面的竖线,这样看起来会显得比较整齐
val line =
    """
      |0: 字符串类型
      |1: 整型
      |2: 浮点型
      |3: User对象类型
      |
      |请选择:
    """.stripMargin

println(line)
val select = StdIn.readInt()
val selectedValue = select match {
    case 0 => "hello"
    case 1 => 1
    case 2 => 2.0
    case _ => new User("user1")
}
selectedValue match {
    case x: Int => println("Int " + x)
    case y: Double => println("Double " + y)
    case z: String => println("String " + z)
    case _ => throw new Exception("not match exception")
}
```

3.2.4 匹配集合

匹配数组的方法如下。

```
val arr = Array(1, 3, 5)
arr match {
    case Array(1, x, y) => println(x + " " + y)
    case Array(0) => println("only 0")
```

```
    case Array(0, _*) => println("0 ...")
    case _ => println("something else")
}
```

匹配列表的方法如下。

```
val lst = List(3, -1)
lst match {
    case 0 :: Nil => println("only 0")
    case x :: y :: Nil => println(s"x: $x y: $y")
    case 0 :: tail => println("0 ...")
    case _ => println("something else")
}
```

匹配元组的方法如下。

```
val tup = (1, 3, 7) tup match {
 case (1, x, y) => println(s"1, $x , $y")
 case (_, z, 5) => println(z)
 case _ => println("else")
}
```

3.2.5　变量声明中的模式匹配

在定义变量的时候，可以使用模式匹配快速获取数据。

示例：获取数组中的元素。

```
val range= Range(0, 10).toArray
range.foreach(println(_))
// 使用模式匹配，获取第二个、第三个、第四个元素的值
val Array(_, x, y, z, _*) = arr
println(s"x=$x, y=$y, z=$z, ")
```

示例：获取列表中的元素。

```
val list = Range(0, 10).toList
// 匹配列表中的第一个、第二个元素的值
val x::y::tail = list
println(s"x=$x, y=$y")
```

3.2.6 匹配样例类

Scala 可以使用模式匹配来匹配样例类，从而可以快速获取样例类中的成员变量。

```
// 定义样例类
case class Action2(count: Long)
case object Action3
val comment1= Action1("参数1")
val comment2= Action2(1000)
val comment3= Action3
val list = List(comment1, comment2, comment3)
list(2) match {
 case Action1(action) => println(s"action=$action")
 case Action2(count) => println(s"count=$count")
 case Action3=> println("run action3")
}
```

使用 @ 符号分隔 case 语句，可以获取用于匹配的整个样例对象。

```
list(0) match {
 // obj 表示获取用于匹配的样例对象，而action 表示获取样例中的元素
 case obj @ Action1(action) => println(s"action=$action");
                               println(s"样例类:$obj")
 case Action2(count) =>  println(s"count=$count")
 case Action3=>  println("run action3")
}
```

3.3 Option 类型

Scala 中的 Option 类型用来表示可选值。这种类型的数据有两种形式：Some(x) 表示实际的值，None 表示没有值。

使用 Option 类型可以有效避免空引用（null）异常。也就是说，将来我们返回某些数据时，可以返回 Option 类型来替代。

```
/**
 * 定义除法操作
```

```
 * @param a: 参数1
 * @param b: 参数2，作为分母
 * @return Option 包装 Double 类型
 */
def dvi(a:Double, b:Double):Option[Double] = {
    if(b != 0) {
        Some(a / b)
    }
    else {
        None
    }
}
def main(args: Array[String]): Unit = {
    val res= dvi(10, 5)
    res match {
        case Some(x) => println(x)
        case None => println("分母为0")
    }
}
```

使用 getOrElse 方法，当 Option 对应的数据是 None 时，可以指定一个默认值，从而避免空指针异常。

```
val res = dvi(10, 0)
println(res.getOrElse("分母为0"))
```

Scala 鼓励使用 Option 类型来封装数据，可以有效减少在代码中判断某个值为 null 的情况。可以使用 getOrElse 方法来针对 None 返回一个默认值。

3.4 偏函数

被包在花括号内没有 match 的一组 case 语句是一个偏函数，它是 PartialFunction[A, B] 的一个实例，A 代表输入参数类型，B 代表返回值类型。

可以这样理解：偏函数是包含一个参数和一个返回值的函数。

```
// func1 是一个输入参数为 Int 类型、返回值为 String 类型的偏函数
val func1: PartialFunction[Int, String] = {
 case 1 => "one"
 case 2 => "two"
```

```
    case 3 => "three"
    case _ => "other" }
println(func1(2))
```

示例：获取列表中能够被 2 整除的数字。

```
val list = List(1,2,3,4,5,6,7,8,9,10)
val tmp_list= list.filter{
    case x if x % 2 == 0 => true
    case _ => false
}
println(tmp_list)
```

3.5 正则表达式

在 Scala 中，可以很方便地使用正则表达式来匹配数据。

Scala 提供了 Regex 类来定义正则表达式。要构造一个 Regex 对象，直接使用 String 类的 r 方法即可。建议使用 3 个双引号来表示正则表达式，不然就要对正则表达式中的反斜线进行转义。其中，圆括号中间的内容可以作为返回值获取。

```
val regEx = """正则表达式""".r
```

示例：检查是否匹配正则表达式。

```
val emailRE = """.+@(.+)\..+""".r
val emailList = List("1234567@email.com", "abcdefg@gmail.com", "one@163.com", "two.com",".aaa.abc.ccc")
// 检查邮箱地址是否匹配正则表达式
val size = emailRE.findAllMatchIn(emailList(0)).size
// 如果匹配则size 为 1，否则 size 为 0
println(size)
```

示例：找出列表中所有不合法的邮箱地址。

```
// 找出列表中所有不合法的邮箱地址
println("不合法的邮箱地址为: ")
emailList.filter{
    eml => emailRE.findAllIn(eml).size < 1
}.foreach(println(_))
```

示例：使用正则表达式进行模式匹配，获取正则表达式中匹配的分组。

```
// 找到所有正确邮箱地址的归属
emailList.foreach {
 case email @ emailRE(belong) => println(s"$email => ${belong}")
 case _ => println("未知")
}
```

3.6 异常处理

我们来看看下面一段代码。

```
def main(args: Array[String]): Unit = {
    val i = 10 / 0
    println("i="+i)
}
Exception in thread "main" java.lang.ArithmeticException: / by zero
at ForDemo$.main(ForDemo.scala:3)
at ForDemo.main(ForDemo.scala)
```

执行代码,可以看到 Scala 抛出了异常,而且没有打印算术值,说明代码出现错误后就终止了。怎么解决该问题呢?

3.6.1 捕获异常

在 Scala 中,可以使用异常处理来解决这个问题。以下为 Scala 中 try...catch...finally 异常处理的语法格式:

```
try {
 // 代码
}
catch {
 case ex1:异常类型1 => 异常处理代码1
 case ex2:异常类型2 => 异常处理代码2
}
finally {
 // 代码
}
```

try 中的代码是我们编写的业务处理代码，catch 中的代码是出现某个异常时需要执行的代码，finally 中的代码是不管是否出现异常都会执行的代码。

```
try {
  val i = 10 / 0
  println("i="+i)
} catch {
  case ex: Exception => println(ex.getMessage)
} finally {
  println("必须执行的代码块")
}
```

3.6.2 抛出异常

我们也可以在一个方法中抛出异常，语法格式和 Java 中的类似，使用 throw new Exception( )。

```
def main(args: Array[String]): Unit = {
    throw new Exception("这是一个异常")
} Exception in thread "main" java.lang.Exception: 这是一个异常
at ForDemo$.main(ForDemo.scala:3)
at ForDemo.main(ForDemo.scala)
```

可以看到，Scala 不需要在 main 方法上声明要抛出的异常，它已经解决了在 Java 中被认为设计失败的检查型异常。下面是 Java 代码。

```
public static void main(String[] args) throws Exception {
    throw new Exception("这是一个异常");
}
```

在 Scala 中也使用 try...catch...finally 处理异常。所有异常处理均在 catch 语句中。异常处理写成如下形式：

```
case ex1:异常类型1 => 异常处理代码1
case ex2:异常类型2 => 异常处理代码2
case ex3:异常类型3 => 异常处理代码3
```

抛出异常使用 throw。Scala 中方法抛出异常不需要像 Java 一样编写异常声明。

3.7 提取器

我们之前已经介绍过 Scala 非常强大的模式匹配功能，通过模式匹配，我们可以快速匹配样例类中的成员变量。

```scala
// 定义样例类
case class Action2(count: Long)
case object Action3
val comment1= Action1("参数1")
val comment2= Action2(1000)
val comment3= Action3
val list = List(comment1, comment2, comment3)
list(2) match {
 case Action1(action) => println(s"action=$action")
 case Action2(count) => println(s"count=$count")
 case Action3=> println("run action3")
}
```

那么是不是所有的类都可以进行这样的模式匹配呢？答案为不是。要支持模式匹配，必须实现一个提取器。

定义提取器

之前我们学习过了，实现一个类的伴生对象中的 apply 方法，可以用类名来快速构建一个对象。伴生对象中，还有一个 unapply 方法。与 apply 方法相反，unapply 方法可将该类的对象拆解为一个一个的元素。要实现一个类的提取器，只需要在该类的伴生对象中实现一个 unapply 方法即可。

示例：实现一个类的提取器，并使用 match 表达式进行模式匹配，提取类中的字段。

```scala
class Student {
    var studentName:String = _    // 姓名
    var studentAge:Int = _    // 年龄

    // 实现一个辅助构造器
    def this(name:String, age:Int) = {
```

```
    this()

    this.studentName= studentName
    this.studentAge= studentAge
  }
}
object Student {
    def apply(studentName:String, studentAge:Int): Student = new Student(studentName, studentAge)
    // 实现一个提取器
    def unapply(arg: Student): Option[(String, Int)] = Some((arg.studentName, arg.studentAge))
}
object extractorDemo{
    def main(args: Array[String]): Unit = {
        val student1= Student("student1", 20)
        student1 match {
            case Student(studentName, studentAge) => println(s"姓名:$studentName 年龄:$studentAge")
            case _ => println("未匹配")
        }
    }
}
```

样例类自动实现了 apply、unapply 方法（可以使用由 scala 社区贡献的 Scalap 工具来反编译一个样例类的字节码）。

3.8 泛型

Scala 和 Java 一样，类、特质、方法都可以支持泛型。我们在学习集合的时候，一般会涉及泛型。

```
scala> val list1:List[String] = List("1", "2", "3")
list1: List[String] = List(1, 2, 3)
scala> val list1:List[String] = List("1", "2", "3")
list1: List[String] = List(1, 2, 3)
```

在 Scala 中，使用方括号来定义类型参数。

3.8.1 定义泛型方法

我们经常会遇到这样的需求：用一个方法来获取任意类型数组的中间的元素。而实现该需求有两种方法：不考虑泛型直接实现（基于 Array[Int] 实现）；加入泛型支持。

示例：不考虑泛型直接实现。

```
def newArr(arr:Array[Int]) = arr(arr.length / 2)
    def main(args: Array[String]): Unit = {
        val arr1 = Array(1,2,3,4,5)
        println(newArr(arr1))
 }
```

示例：加入泛型支持。

```
def newArr[A](arr:Array[A]) = arr(arr.length / 2)
def main(args: Array[String]): Unit = {
    val arr1 = Array(1,2,3,4,5)
    val arr2 = Array("a", "b", "c", "d", "f")
    println(newArr[Int](arr1))
    println(newArr[String](arr2))
    // 简写方式
    println(newArr(arr1))
    println(newArr(arr2))
}
```

3.8.2 定义泛型类

接下来我们通过实现一个 Pair 类（一对数据）来讲解 Scala 泛型相关的知识点。Pair 类包含两个值，而且两个值的类型不固定。

```
// 类名后面的方括号，表示这个类可以使用两个类型，分别是 T 和 S
// 这个名字可以任意取
class Pair[T, S](val first: T, val second: S)
case class User(var name:String, val age:Int)
object Pair {
    def main(args: Array[String]): Unit = {
```

```
        val p1 = new Pair[String, Int]("username1", 20)
        val p2 = new Pair[String, String]("username2", "2020-02-19")
    val p3 = new Pair[User, User](User("username3", 20), User("username4", 30))
  }
}
```

要定义一个泛型类，可以直接在类名后面加上方括号，指定要使用的类型参数。上述的 T、S 都是类型参数，代表一个类型。指定了类对应的类型参数后，就可以使用这些类型参数来定义变量。

3.8.3 上下界

在 Pair 类中，我们只想保存 User 类的对象，因为我们要添加一个方法，让用户之间能够通信。

```
def chatting(msg:String) = println(s"${first.name}对${second.name}说：$msg")
```

但因为 Pair 类中根本就不知道 first 有 name 这个字段，上述代码会出现编译错误。而且添加了这个方法，就表示 Pair 类现在只能支持 User 类或者 User 的子类的泛型。所以，我们需要给 Pair 的泛型参数添加一个上界。

使用 <: 类型名给类型添加一个上界，表示泛型参数必须从上界继承。

```
// 类名后面的方括号，表示这个类可以使用两个类型，分别是 T 和 S
// 这个名字可以任意取
class Pair[T <: User, S <:User](val first: T, val second: S) {
 def chatting(msg:String) =
    println(s"${first.name}对${second.name}说：$msg")
}
class User(var name:String, val age:Int)
object Pair {
    def main(args: Array[String]): Unit = {
        val p= new Pair(new User("user1", 20), new User("user2", 25)
        p.chatting("hello! ")
    }
}
```

User 类有 2 个子类：Student、Teacher。要控制 User 对象只能和 User 对象、Student 对象聊天，但是不能和 Teacher 对象聊天，就需要给泛型参数添加一个下界。

```
// 类名后面的方括号，表示这个类可以使用两个类型，分别是 T 和 S
// 这个名字可以任意取
class Pair[T <: User, S >: Student<:User](val first: T, val second: S) {
    def chatting(msg:String) =
        println(s"${first.name}对${second.name}说：$msg")
}
class User(var name:String, val age:Int)
class Teacher(name:String) extends User(name, age)

class Student(name:String, age:Int) extends Teacher(name, age)
object Pair {
    def main(args: Array[String]): Unit = {
    // 编译错误：第二个参数必须是 User 的子类（包括本身）、Student 的父类（包括本身）
        val p3 = new Pair(new User("user", 20), new Teacher("student"))
        p3.chat("hello! ")
    }
}
```

U >: T表示U必须是类型T的父类或本身。S <: T表示S必须是类型T的子类或本身。

3.8.4 非变、协变和逆变

下面是一个类型转换的示例。

```
class Pair[T]
object Pair {
    def main(args: Array[String]): Unit = {
        val p1 = Pair("hello")
        // 编译报错，无法将 p1 转换为 p2
        val p2:Pair[AnyRef] = p1
        println(p2)
    }
}
```

1. 非变

class Pair[T]{}，这种情况就是非变（默认）。类型 B 是 A 的子类型，Pair[A] 和 Pair[B] 没有任何从属关系。这种情况和在 Java 中的是一样的。

2. 协变

class Pair[+T]，这种情况是协变。类型 B 是 A 的子类型，Pair[B] 可以认为是 Pair[A] 的子类型。这种情况下，参数化类型的方向和类型的方向是一致的。

3. 逆变

class Pair[-T]，这种情况是逆变。类型 B 是 A 的子类型，Pair[A] 可以认为是 Pair[B] 的子类型。这种情况下，参数化类型的方向和类型的方向是相反的。

```scala
class Super
class Sub extends Super
//非变
class Temp1[A](title: String)
//协变
class Temp2[+A](title: String)
//逆变
class Temp3[-A](title: String)
object Demo {
  def main(args: Array[String]): Unit = {
    val a = new Sub()
    // 没有问题，Sub 是 Super 的子类
    val b:Super = a
    // 非变
    val t1:Temp1[Sub] = new Temp1[Sub]("test")
    // 报错！默认不允许转换
    // val t2:Temp1[Super] = t1
    // 协变
    val t3:Temp2[Sub] = new Temp2[Sub]("test")
    val t4:Temp2[Super] = t3

    // 非变
    val t5:Temp3[Super] = new Temp3[Super]("test")
    val t6:Temp3[Sub] = t5
  } }
```

3.9 Actor

Actor 并发编程模型是一种基于事件模型的并发机制。下面通过与 Java 并发编程模

型对比来学习 Scala 编程语言中的 Actor。

3.9.1 Java 并发编程的问题

在 Java 并发编程中，每个对象都有一个逻辑监视器（monitor），可以控制对象的多线程访问。我们添加 synchronized 关键字来标识，需要进行同步加锁访问。这样，通过加锁的机制来确保同一时间只有一个线程访问共享数据。但这种方式存在资源争夺（见图 3-1）以及死锁问题（见图 3-2）。程序越复杂，解决起来越麻烦。

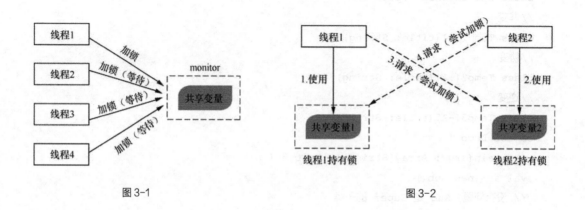

图 3-1　　　　　　　　　　　　图 3-2

3.9.2 Actor 并发编程模型

Actor 是 Scala 提供的一种与 Java 并发编程模型完全不一样的并发编程模型，是一种基于事件模型的并发机制。Actor 并发编程模型是一种不共享数据、依赖消息传递的并发编程模式，可以有效避免资源争夺、死锁等情况，如图 3-3 所示。

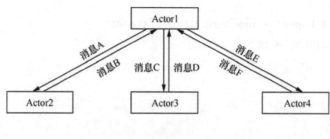

图 3-3

3.9.3　Java 并发编程与 Actor 并发编程

Java 并发编程与 Actor 并发编程对比如表 3-1 所示。

表 3-1

Java 并发编程	Actor 并发编程
"共享数据 – 锁"模型	
每个对象都有一个 monitor，用于监视线程对共享数据的访问	不共享数据，Actor 之间通过 Message 通信
加锁代码使用 synchronized 标识	
死锁问题	
每个线程内部是顺序执行的	每个 Actor 内部是顺序执行的

Scala 在 2.11.x 版本中加入了 Akka 并发编程框架。Actor 并发编程模型和 Akka 很像，我们这里学习 Actor 是为学习 Akka 并发编程框架做准备。

3.10　Actor 编程案例

下面我们通过 Actor 编程案例来体验 Actor 并发编程模型。

3.10.1　创建 Actor

创建 Actor 的方式和 Java 中创建线程的方式很类似，具体步骤如下。

① 定义类或对象继承 Actor 特质。

② 实现 act 方法。

③ 调用 start 方法执行 Actor。

示例：创建两个 Actor，一个 Actor 打印 1 ~ 10，另一个 Actor 打印 11 ~ 20。

```
// 继承 Actor 特质
object Actor1 extends Actor {
```

```
    // 实现 act 方法
        override def act(): Unit =
            (1 to 10).foreach{
                num => println(s"线程1: $num")
            }
    }
    object Actor2 extends Actor {
      override def act(): Unit =
          (11 to 20).foreach{
              num =>    println(s"线程2: $num")
          }
    }
    object ActorDemo {
        def main(args: Array[String]): Unit = {
        // 调用 start 方法执行 Actor
          Actor1.start()
          Actor2.start()
      }
    }
```

上述代码分别调用了单例对象的 start 方法（对象本质也是一个类，可看成包含静态成员的类），会在 JVM 中开启两个线程来执行 act 方法。

Actor 的执行顺序如下。

① 调用 start 方法执行 Actor。

② 自动执行 act 方法。

③ 向 Actor 发送消息。

④ act 方法执行完成后，Actor 程序会自动调用 Actor 对象中自带的 exit 方法。

注意：Actor 是并发执行的；act 方法执行完成后，Actor 程序就退出了。

3.10.2 发送消息 / 接收消息

Actor 是基于事件（消息）的并发编程模型，那么 Actor 是如何发送消息和接收消息的呢？

我们可以使用 3 种方式来发送消息。

- 发送异步消息，没有返回值。
- 发送同步消息，等待返回值。
- 发送异步消息，返回值是 Future[Any]。

语法格式如下。

```
actor ! 消息
```

示例：创建两个 Actor，Actor1 发送一个异步字符串消息给 Actor2，Actor2 接收该消息后将其打印。

```scala
object ActorSender extends Actor {
    override def act(): Unit = {
        // 发送字符串消息给 Actor2
        val msg = "hello,ActorSender"
        println(s"ActorSender: 发送消息$msg")
        ActorReceiver ! msg
    }
}
object ActorReceiver extends Actor {
 override def act(): Unit =
    receive {
        case msg: String => println(s"接收 Actor: 接收到$msg")
    }
}
object ActorMsgDemo {
    def main(args: Array[String]): Unit = {
        ActorSender.start()
        ActorReceiver.start()
    }
}
```

3.10.3 持续接收消息

上面示例中，ActorReceiver 调用 receive 来接收消息，但接收一次后，Actor 就退出了。

```scala
object ActorSender extends Actor {
    override def act(): Unit = {
        // 发送字符串消息给 Actor2
        val msg = "hello,ActorSender"
```

```
            println(s"ActorSender:发送消息$msg")
            ActorReceiver ! msg
            // 再次发送一条消息,ActorReceiver 无法接收到
            ActorReceiver ! "this is second test message"
        }
    }
    object ActorReceiver extends Actor {
        override def act(): Unit =
            receive {
                case msg: String => println(s"接收 Actor:接收到$msg")
            }
    }
    object ActorMsgDemo {
        def main(args: Array[String]): Unit = {
            ActorSender.start()
            ActorReceiver.start()
        }
    }
```

上述代码中,ActorReceiver 无法接收 ActorSender 发送的第二条消息。我们希望 ActorReceiver 能够接收多条消息,怎么实现呢?只需要使用一个 while 循环,不停地调用 receive 来接收消息。

```
    object ActorSender extends Actor {
        override def act(): Unit = {
            // 发送字符串消息给 Actor2
            val msg = "hello,ActorSender"
            println(s"ActorSender:发送消息$msg")
            ActorReceiver ! msg
            // 再次发送一条消息,ActorReceiver 无法接收到
            ActorReceiver ! "this is second test message"
        }
    }
    object ActorReceiver extends Actor {
        override def act(): Unit =
        // 使用 while 循环不停地接收消息
        while(true) {
            receive {
                case msg: String => println(s"接收 Actor:接收到$msg")
            }
```

```
        }
    }
object ActorMsgDemo {
    def main(args: Array[String]): Unit = {
        ActorSender.start()
        ActorReceiver.start()
    }
}
```

3.10.4 共享线程

上述代码使用 while 循环来不断接收消息。但如果当前 Actor 没有接收到消息，线程就会处于阻塞状态。考虑到如果有很多的 Actor，就有可能导致很多线程都处于阻塞状态。当有新的消息来时，需要重新创建线程来处理，这样会导致频繁地进行线程创建、销毁和切换，从而影响运行效率。在 Scala 中，可以使用 loop+react 来复用线程，这比 while +receive 更高效。

示例：使用 loop + react 重写上述示例。

```
loop {
    react{
        case msg:String => println(s"接收 Actor: 接收到$msg")
    }
}
```

3.10.5 发送和接收自定义消息

我们前面发送的消息是字符串类型的，Actor 中也支持发送自定义消息。

示例：每一个消息使用样例类来封装，每一个消息都有一个 id:Int、msg:String 成员。

```
object MessageActor extends Actor {
    override def act(): Unit =
        loop {
            react{
                // 接收同步消息
                case SyncMessage(id, msg) =>
                    println(s"接收到 SyncMessage 消息:id=${id}, msg=${msg}")
```

```scala
                // 使用 sender 来获取发送方 Actor 的引用
                sender ! ReplyMessage(2, "回复消息")
            case ASyncWithoutMessage(id, msg) =>
                println(s"接收到 ASyncWithoutMessage 消息
                : id=${id}, msg=${msg}")
            case ASyncWithMessage(id, msg) =>
                println(s"接收到 ASyncWithoutMessage 消息: id=${id}, msg=${msg}")
                // 10s后再回复消息
                TimeUnit.SECONDS.sleep(10)
                sender ! ReplyMessage(5, "回复消息")
            case _ => println("未知消息")
        }
    }
}

object MainActor {
    def main(args: Array[String]): Unit = {
        MessageActor.start()
        // 发送同步消息
        println("发送同步消息")
        println("-----")
        val replyMessage = MessageActor !? SyncMessage(1, "同步消息")
        println(replyMessage)
        println("-----")
        // 发送异步无返回消息
        println("发送异步无返回消息")
        MessageActor ! ASyncWithoutMessage(3, "异步无返回消息")
        // 睡眠10s，等待 Actor 接收并打印异步消息，方便观察测试
        TimeUnit.SECONDS.sleep(10)
        println("-----")
        // 发送异步有返回消息
        println("发送异步有返回消息")
        val future: Future[Any] = MessageActor !! ASyncWithMessage(4, "异步有返回消息")
        // Future 表示对返回消息的封装，因为发送的是异步消息，所以不确定在将来哪个时间会返回消息
        // 使用while循环，不断调用 isSet 来检查是否已经接收到消息
        while (!future.isSet) {
        }
        val asyncReplyMessage = future.apply()
        println(s"接收到 ReplyMessage: $asyncReplyMessage")
    }
}
```

在编写 Actor 程序时，一般使用样例类来封装消息。在 Actor 的 act 方法中，可以使用 sender 来获取发送方 Actor 的引用。Future 表示对有返回的异步消息的封装，虽然获取到了 Future 的返回值，但 Future 中不一定有值，因为可能在将来的某一时刻才会返回消息。使用 Future 的 isSet 可以检查是否已经收到返回消息，使用 apply 方法可以获取返回的消息。使用 TimeUnit.SECONDS.sleep 来让 Actor 对应的线程进入睡眠状态。

3.10.6 基于 Actor 实现 WordCount 案例

接下来，我们使用 Actor 并发编程模型实现多文件的单词统计（WordCount），实现步骤如下。

① MainActor 获取要进行单词统计的文件。

② 根据文件数量创建对应的 WordCountActor。

③ 将文件名封装为消息并发送给 WordCountActor。

④ WordCountActor 接收消息，并统计单个文件的单词数量。

⑤ 将单词计数结果回复给 MainActor。

⑥ MainActor 等待所有的 WordCountActor 都已经成功返回消息，然后进行结果合并。

```scala
case class WordCountTask(fileName:String)
case class WordCountResultMessage(wcResult:Map[String,Int])
class WordCountActor extends Actor {
    override def act(): Unit = {
    loop {
        react {
            // 接收任务，获取文件名
            case WordCountTask(fileName) =>
            val actorId = this.toString.split("@")(1)
            println(s"${actorId}接收到任务：$fileName")
            // 读取文件
            val lineArr: Array[String] =
            Source.fromFile(fileName).mkString.split("\r\n")
            val wordArr: Array[String] = lineArr.flatMap(_.split(" "))
            val wordnumArr: Array[(String, Int)] = wordArr.map(_->1)
            val groupedWordNumMap: Map[String, Array[(String, Int)]] =
```

```scala
            wordnumArr.groupBy(_._1)
                        val result: Map[String, Int] =
            groupedWordNumMap.mapValues(_.foldLeft(0)(_ + _._2))
                        println(s"${actorId}返回结果:${result.toList.sorted}")
                        sender ! WordCountResultMessage(result)
                }
            }
    }
}
object WordCountDemo {
            def main(args: Array[String]): Unit = {
                val dir = "./data/wordTestFile"
                val files = List(s"$dir/w1.txt", s"$dir/w2.txt", s"$dir/w3.txt")
                // 启动若干个 WordCountActor
                val wordCountActors: List[(String, WordCountActor)] =
                    files.map {
                        file =>
                            val wordCountActor = new WordCountActor()
                            wordCountActor.start()
                            (file, wordCountActor)
                    }
                // 发送文件名给每一个启动的WordCountActor
                val futureList: List[Future[Any]] = wordCountActors.map {
                    fileAndActor =>
                        // 发送 WordCountTask 给WordCountActor
                        fileAndActor._2 !! WordCountTask(fileAndActor._1)
                }
                // 等待所有结果返回
                while(futureList.count(!_.isSet) != 0) {}
                // 获取所有结果
                val wordCountResults: List[(String, Int)] =
futureList.flatMap(_.apply().asInstanceOf[WordCountResultMessage].wcResult.toList)
                // 按照单词进行分组
                val groupWordCountResults: Map[String, List[(String, Int)]] =
wordCountResults.groupBy(_._1)
                val finalResult: Map[String, Int] =
groupWordCountResults.mapValues(_.foldLeft(0)(_ + _._2))
                println("-" * 10)
                println("合并后的结果为: ")
            println(finalResult.toList.sorted)
            }
}
```

3.11 本章总结

本章主要介绍了样例类、模式匹配、偏函数以及正则表达式的使用方法，讲解了非变、协变、逆变和上下界的相关概念。本章也详细介绍了 Actor 并发编程模型，从逻辑和思路方面进行了阐述，对 Scala 的程序开发有很大的拓展。

3.12 本章习题

1. 通过代码实现样例类、模式匹配、偏函数以及正则表达式。
2. 熟练掌握 Actor 并发编程模型。

第 4 章

Scala 函数式编程思想

Scala 混合了面向对象和函数式的特性，在函数式编程语言中，函数和 Int、String 等其他类型处于同等的地位，可以像其他类型的变量一样被传递和操作。本章重点介绍作为值的函数、匿名函数、柯里化、闭包、隐式转换和隐式参数等。

4.1 作为值的函数

在 Scala 中，函数和数字、字符串类似，可以将函数传递给一个方法。我们可以对算法进行封装，然后将具体的动作传递给方法，这种特性很有用。

我们之前学习的 map 方法，就可以接收一个函数，实现列表的转化。

将列表中的每一个元素都转换为 # 的步骤如下。

① 创建一个函数，用于将数字转换为指定个数的 #。

② 创建一个列表，调用 map 方法。

③ 打印转换后的列表。

代码如下所示。

```
val func: Int => String = (num:Int) =>
println((1 to 10).map(func))
```

4.2 匿名函数

在 Scala 中，可以不用给函数赋值，没有赋值的函数就是匿名函数。Scala 中定义匿名函数的方法很简单，箭头左边是参数列表，右边是函数体。使用匿名函数后，代码将会变得很简洁。下面就定义了一个接收 Int 类型输入参数的匿名函数：

```
var inc = (x:Int) => x+1
```

上述定义的匿名函数，其实是下面这种写法的简写形式：

```
def add = new Function1[Int, Int]{
    def apply(x:Int) = x+1
}
```

假如给 (num:Int) => "*" * num 函数赋值了一个变量：

```
val list = List(1,2,3,4)
//表示生成指定数量的字符串"*"，如果num=5，那么就会生成"*****"字符串
val func_num = (num:Int)=> "*" * num
println(list.map(func_num))
```

但是上面这种写法有一些累赘，利用 Scala 的匿名函数可以简化上述代码，可以不用给函数赋值，直接用匿名函数的方式进行传参。优化代码如下：

```
println(list.map(num => "*" * num))
/* 因为此处num变量只使用了一次，而且只是进行简单的计算，所以可以省略参数列表，使用"_"替代参数*/
println(list.map("*" * _))
```

4.3 柯里化

在 Scala 和 Spark 的源码中，大量使用了柯里化。为了后续能够读懂源码，我们需要来了解柯里化的知识。

柯里化（Currying）是指将原先接收多个参数的方法转换为多个只有一个参数的参数列表的过程。

例如，一个方法 func 如下：

```
def func(x:Int,y:Int) = x + y
```

柯里化后：

```
def func(x:Int)(y:Int) = x + y
```

柯里化解析：

```
    def func(x:Int) = {
(y:Int) = x + y
}
```

def func(x:Int) 返回的是一个方法，def func(x:Int)(y:Int) 调用返回的方法返回最终的值，经过柯里化之后，转化为一个单参数的列表。

示例：编写一个方法，用来完成两个 Double 类型数字的计算，包含加、减、乘、除。

```
object Test {
  def calc_math(x: Double, y: Double)(func_calc: (Double, Double) => Double) = {
    func_calc(x, y)
  }

  def main(args: Array[String]): Unit = {
    println(calc_math(10, 10) {
      (x, y) => x + y
    })  // 20
    println(calc_math(10, 10)(_ / _))  // 1
    println(calc_math(10, 10)(_ * _))  // 100
    println(calc_math(10, 10)(_ - _))  // 0
  }
}
```

4.4 闭包

闭包其实就是一个函数，只不过这个函数的返回值依赖于声明在函数外部的变量。可以将其简单地理解为：访问不在当前作用域范围的一个函数。下面用两个示例解释这一点。

示例：定义一个闭包 add 函数，用于将两个数相加。

```
object Test {
// 闭包
```

```
def main(args: Array[String]): Unit = {
val y = 10
val add = (x: Int) => {
  x + y
  }
 println(add(10)) // 结果为20
}
}
```

示例：特殊地，柯里化是一个闭包。

```
def add(x: Int)(y: Int) = {
x + y
}
```

等价于：

```
def add(x: Int) = {
   (y: Int) => x + y
}
```

4.5 隐式转换

隐式转换是 Scala 中非常有特色的功能，也是 Java 等其他编程语言没有的功能。我们可以很方便地利用隐式转换来丰富现有类的功能。在 Akka 并发编程、Spark SQL、Flink 中都会看到隐式转换的身影。

所谓隐式转换，是指以 **implicit** 关键字声明的带有单个参数的方法。它会被自动调用，将某种类型转换为另外一种类型。使用方法如下。

① 在 object 中定义隐式转换方法（使用 implicit）。

② 在需要用到隐式转换的地方，引入隐式转换（使用 import）。

③ 自动调用隐式转换后的方法。

示例：使用隐式转换输出"xiaoming birthday, eating cake"。

```
class EatCake(person: String) {
   //定义吃生日蛋糕的方法
   def eat() = {
     println(person + " birthday , eating cake")
```

```scala
    }
}

object EatCake {
  //定义隐式转换方法
  implicit def birthday(person: String) = new EatCake(person)
}


object Test {
  def main(args: Array[String]): Unit = {
    // 引入隐式转换
    import EatCake.birthday
    println("xiaoming".eat()) //输出"xiaoming birthday，eating cake"
  }
}
```

当对象调用类中不存在的方法或者成员时，编译器会自动将对象进行隐式转换。当方法中的参数的类型与目标类型不一致时，会自动引入隐式转换。

前面，我们手动使用了 import 来引入隐式转换。是否可以不使用 import 呢？ 在 Scala 中，如果在当前作用域中有隐式转换方法，会自动引入隐式转换。

示例：将隐式转换方法定义在 main 所在的 object 中。

```scala
class EatCake(person: String) {
  //定义吃生日蛋糕的方法
  def eat() = {
    println(person + " birthday, eating cake")
  }
}


object Test {
  def main(args: Array[String]): Unit = {
    // 引入隐式转换
    // import EatCake.birthday
    // 替换成下面隐式转换的方式
    implicit def birthday(person: String) = new EatCake(person)
    // 同样调用 EatCake 中的eat方法
    println("xiaoming".eat()) //输出"xiaoming birthday, eating cake"
  }
```

4.6 隐式参数

方法可以带一个标记为 **implicit** 的参数列表。这种情况下，编译器会查找默认值，提供给该方法。

在方法后面添加一个参数列表，参数列表使用 implicit 修饰。在 object 中定义 implicit 修饰的隐式值。即使不传入 implicit 修饰的参数列表，编译器也会自动查找默认值。

注意：与隐式转换一致，可以使用 import 引入隐式参数，且如果在当前作用域定义了隐式值，那么参数会自动引入。当然，标记为隐式参数的，我们也可以手动为该参数添加默认值。

```
def fun(n: Int)(implicit t1: String, t2: Double = 1.1)
```

示例：为书名添加方头括号。

```scala
//使用implicit定义一个参数
def quote(bookname: String)(implicit delimiter: (String, String)) = {
  delimiter._1 + bookname + delimiter._2
}

//隐式参数
object ImplicitParam {
  implicit val DELIMITERS = ("【", "】")
}

def main(args: Array[String]): Unit = {
  //引入隐式参数
  import ImplicitParam.DELIMITERS
  println(quote("bookName")) // 【bookName】
}
```

4.7 Akka 并发编程框架

Akka 是一个用于构建高并发、分布式和可扩展的基于事件驱动的应用的工具包。可以同时使用 Scala 和 Java 语言来开发基于 Akka 的应用程序。

4.7.1　Akka 特性

Akka 有以下 4 个特性。

- 提供基于异步非阻塞、高性能的事件驱动编程模型。
- 内置容错机制，允许 Actor 在出错时进行恢复或者重置操作。
- 超级轻量级的事件处理（每 GB 堆中存储几百万 Actor）。
- 使用 Akka 可以在单机上构建高并发程序，也可以在网络中构建分布式程序。

4.7.2　Akka 通信过程

Akka 中的角色如下。

- ProducerActor（StudentActor）。
- ConsumerActor（TeacherActor，oneceive 方法接收消息）。
- ActorRef（tell 方法，发送消息给 MessageDispatcher）。
- ActorSystem（actorOf 方法，创建 ActorRef，ActorRef 就是 ConsumerActor 的 Proxy）。
- MailBox。
- Dispatcher。
- Message。

Akka 工作流程如下。

① ProducerActor（StudentActor）创建 ActorSystem。

② 通过 ActorSystem 创建一个叫 ActorRef 的对象（ActorRef 是 TeacherActor 的一个代理），QuoteRequest 消息就是发送给 ActorRef 的。

③ ActorRef 将消息发送给 MessageDispatcher。

④ Dispatcher 将消息投递到目标 Actor 的 MailBox 中。

⑤ Dispatcher 将 MailBox 放到一个线程中去执行。

⑥ MailBox 将消息出队并最终将其委托给真实的 TeacherActor 的接收方法去处理。

4.7.3 创建 ActorSystem

ActorSystem 是进入 Actor 世界的一扇大门。通过它你可以创建或终止 Actor，甚至可以关闭整个 Actor 环境。

Actor 是一个分层的结构，ActorSystem 类似于 java.lang.Object 或者 scala.Any 的角色。也就是说，它是所有 Actor 的根对象。当你通过 ActorSystem 的 actorOf 方法创建一个 Actor 时，你其实创建的是 ActorSystem 下面的一个 Actor，它是所有用户创建的 Actor 的父类，如图 4-1 所示。

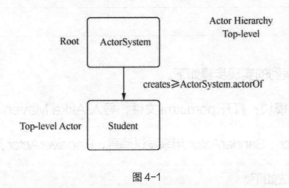

图 4-1

下面通过一个简单的示例深入了解一下 ActorSystem 的通信过程。

① 学生创建一个 ActorSystem。

② 通过 ActorSystem 创建一个 ActorRef（老师的引用），并将消息发送给 ActorRef。

③ ActorRef 将消息发送给 MessageDispatcher（消息分发器）。

④ MessageDispatcher 将消息按照顺序保存到目标 Actor 的 MailBox 中。

⑤ MessageDispatcher 将 MailBox 放到一个线程中去执行。

⑥ MailBox 按照顺序取出消息，最终将消息传递给 TeacherActor 的接收方法去处理。

4.8 Akka 编程入门案例

基于 Akka 创建两个 Actor，Actor 之间可以互相发送消息，如图 4-2 所示。

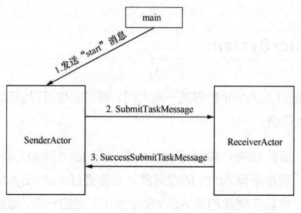

图 4-2

4.8.1 实现步骤

Akka 编程入门案例的实现步骤如下。

① 创建 Maven 模块：打开 pom.xml 文件，导入 Akka Maven 依赖和插件。

② 创建两个 Actor：SenderActor 用来发送消息，ReceiverActor 用来接收、回复消息。

创建 Actor 的方法如下。

① 创建 ActorSystem。

② 创建自定义 Actor。

③ ActorSystem 加载 Actor。

使用样例类封装消息的方法如下。

① SubmitTaskMessage——提交任务消息。

② SuccessSubmitTaskMessage——任务提交成功消息。

③ 使用类似于之前学习的 Actor 方式，使用 ! 发送异步消息。

4.8.2 配置 Maven 模块依赖

创建 Maven 模块的方法如下，打开 Maven 的配置文件 pom.xml，导入 Akka 编程的 Maven 依赖和插件。

```xml
<dependency>
    <groupId>com.typesafe.akka</groupId>
    <artifactId>akka-actor_2.11</artifactId>
    <version>2.3.14</version>
</dependency>
<dependency>
    <groupId>com.typesafe.akka</groupId>
    <artifactId>akka-remote_2.11</artifactId>
    <version>2.3.14</version>
</dependency>
```

实例代码如下。

```scala
import akka.actor.{Actor, ActorRef, ActorSystem, Props}
import com.typesafe.config.ConfigFactory

object Test {
  case class SubmitTaskMessage(msg:String)
  case class SuccessSubmitTaskMessage(msg:String)
  object SenderActor extends Actor {
    override def preStart(): Unit = println("执行发送, begin")
    override def receive: Receive = {
      case "start" =>
        val receiverActor = this.context.actorSelection("/receiverActor")
        receiverActor ! SubmitTaskMessage("第一个任务!")
      case SuccessSubmitTaskMessage(msg) =>
        println(s"接收到来自${sender.path}的消息: $msg")
    }
  }

  object ReceiverActor extends Actor {
    override def preStart(): Unit = println("开始接收消息啦!!")
    override def receive: Receive = {
      case SubmitTaskMessage(msg) =>
        println(s"接收到来自${sender.path}的消息: $msg")
        sender ! SuccessSubmitTaskMessage("完成提交")
      case _ => println("未匹配的消息类型")
    }
  }

  def main(args: Array[String]): Unit = {
    val actorSystem = ActorSystem("SimpleakkaDemo", ConfigFactory.load())
```

```
        val senderActor: ActorRef = actorSystem.actorOf(Props(SenderActor), "senderActor")
        val receiverActor: ActorRef = actorSystem.actorOf(Props(ReceiverActor), "receiverActor")
        senderActor ! "start"
    }
}
```

4.9　Akka 定时任务

如果我们想要使用 Akka 框架在特定时刻执行一些任务，该如何处理呢？Akka 提供了一个 scheduler 对象，用来实现定时调度功能。使用 ActorSystem.scheduler.schedule 方法，可以启动一个定时任务。

schedule 方法针对 Scala 提供两种使用形式。

第一种：

```
def schedule(
    initialDelay: FiniteDuration,        // 延迟多久后启动定时任务
    interval: FiniteDuration,            // 每隔多久执行一次
    receiver: ActorRef,                  // 给哪个 Actor 发送消息
    message: Any)                        // 要发送的消息
    (implicit executor: ExecutionContext)    // 隐式参数：需要手动导入
```

第二种：

```
 def schedule(
    initialDelay: FiniteDuration,        // 延迟多久后启动定时任务
    interval: FiniteDuration             // 每隔多久执行一次
    )(f: ⇒ Unit)                         // 定期要执行的函数，可以将逻辑写在这里
    (implicit executor: ExecutionContext)    // 隐式参数：需要手动导入
```

示例：定义一个 Actor，每秒发送一个消息给 Actor，Actor 收到后打印消息（使用发送消息方式实现）。

```
import akka.actor.{Actor, ActorRef, ActorSystem, Props}
import com.typesafe.config.ConfigFactory

object ScalaTest {
  object ReceiverActor extends Actor {
```

```scala
    override def receive: Receive = {
      case x => println(x)
    }
  }
  def main(args: Array[String]): Unit = {
    val actorSystem = ActorSystem("actor System", ConfigFactory.load())
    val receiverActor = actorSystem.actorOf(Props(ReceiverActor))
    // 启动scheduler，定期发送消息给Actor
    // 导入隐式转换
    import scala.concurrent.duration._
    // 导入隐式参数
    import actorSystem.dispatcher
    actorSystem.scheduler.schedule(0 seconds,
      1 seconds,
      receiverActor, "hello actor")
  }
}
```

示例：定义一个 Actor，每秒发送一个消息给 Actor，Actor 收到后打印消息（使用自定义方式实现）。

```scala
import akka.actor.{Actor, ActorRef, ActorSystem, Props}
import com.typesafe.config.ConfigFactory

object ScalaTest {
  object ReceiverActor extends Actor {
    override def receive: Receive = {
      case x => println(x)
    }
  }
  def main(args: Array[String]): Unit = {
    val actorSystem = ActorSystem("Simple Akka Demo", ConfigFactory.load())
    val senderActor: ActorRef = actorSystem.actorOf(Props(ReceiverActor), "receive Actor")
    import actorSystem.dispatcher
    import scala.concurrent.duration._
    actorSystem.scheduler.schedule(0 seconds, 1 seconds) {
      senderActor ! "timer"
    }
  }
}
```

这里要注意两个问题。

- 需要导入隐式转换（import scala.concurrent.duration._）才能调用0seconds方法。
- 需要导入隐式参数（import actorSystem.dispatcher）才能启动定时任务。

4.10 实现两个进程之间的通信

本节介绍基于Akka实现两个进程之间的通信。WorkerActor启动后连接MasterActor并发送消息，MasterActor接收到消息后再给Worker回复消息，如图4-3所示。

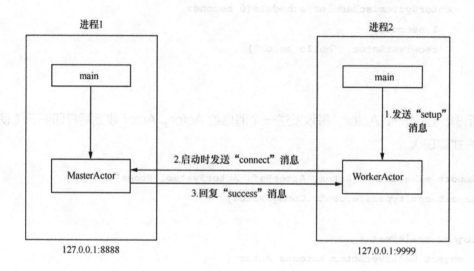

图4-3

Worker实现的步骤如下。

① 创建Maven模块，导入依赖和配置文件application.conf。

② 创建并启动WorkerActor。

③ 发送"setup"消息给WorkerActor，WorkerActor接收并打印消息。

④ 启动测试WorkTest。

```
//创建WorkerActor对象
import akka.actor.Actor
object WorkerActor extends Actor {override def receive: Receive = {
    case "setup" => {
      println("WorkerActor：接收到消息setup")
```

```scala
        // 发送消息给Master
        // 1. 获取到MasterActor的引用
        // MasterActor的引用路径: akka.tcp://actorSystem@127.0.0.1:8888/user/masterActor
        val masterActor = context.actorSelection("akka.tcp://actorSystem@127.0.0.1:8888/user/masterActor")

        // 2. 发送消息给MasterActor
        masterActor ! "connect"
      }
      case "success" => {
        println("WorkerActor：收到success消息")
      }
    }
  }
}
//创建WorkerTest对象
import com.typesafe.config.ConfigFactory
import akka.actor.{Actor, ActorSystem, Props}
import WorkerActor
object WorkerTest {
  def main(args: Array[String]): Unit = {
    // 1. 创建一个ActorSystem
    val actorSystem = ActorSystem("actorSystem", ConfigFactory.load())

    // 2. 加载Actor
    val workerActor = actorSystem.actorOf(Props(WorkerActor), "workerActor")
    // 3. 发送消息给Actor
    workerActor ! "setup"
  }
}
```

```
//配置application.conf文件
akka.actor.provider = "akka.remote.RemoteActorRefProvider"
akka.remote.netty.tcp.hostname = "127.0.0.1"
akka.remote.netty.tcp.port = "1000"
```

```xml
//配置POM文件
<dependency>
    <groupId>com.typesafe.akka</groupId>
    <artifactId>akka-actor_2.11</artifactId>
    <version>2.3.14</version>
</dependency>
```

```xml
<dependency>
    <groupId>com.typesafe.akka</groupId>
    <artifactId>akka-remote_2.11</artifactId>
    <version>2.3.14</version>
</dependency>
```

Master 实现的步骤如下。

① 创建 Maven 模块,导入依赖和配置文件 application.conf。

② 创建并启动 MasterActor。

③ WorkerActor 发送"connect"消息给 MasterActor。

④ MasterActor 回复"success"消息给 WorkerActor。

⑤ WorkerActor 接收并打印消息。

⑥ 启动 Master、Worker 测试:先启动 Master,再启动 Worker。

```scala
//创建MasterActor对象
import akka.actor.Actor
object MasterActor extends Actor {
  override def receive: Receive = {
    case "connect" => {
      println("MasterActor: 接收到connect消息")
      //获取发送者Actor的引用
      sender !"success"
    }
  }
}
//配置application.conf文件
akka.actor.provider = "akka.remote.RemoteActorRefProvider"
akka.remote.netty.tcp.hostname = "127.0.0.1"
akka.remote.netty.tcp.port = "2000"

//创建ScalaTest对象
import akka.actor.{Actor, ActorRef, ActorSystem, Props}
import com.typesafe.config.ConfigFactory
object ScalaTest {
  def main(args: Array[String]): Unit = {
    val actorSystem = ActorSystem("actor System", ConfigFactory.load())
    //加载Actor
```

```
        val masterActor = actorSystem.actorOf(Props(MasterActor), "masterActor")
    }
}
//配置POM文件
<dependency>
    <groupId>com.typesafe.akka</groupId>
    <artifactId>akka-actor_2.11</artifactId>
    <version>2.3.14</version>
</dependency>
<dependency>
    <groupId>com.typesafe.akka</groupId>
    <artifactId>akka-remote_2.11</artifactId>
    <version>2.3.14</version>
</dependency>
```

4.11 本章总结

本章主要介绍了 Scala 高阶函数的概念（作为值的函数、匿名函数、柯里化、闭包等）。通过本章的学习，读者应充分掌握 Scala 函数式编程思想，重点掌握隐式转换和隐式参数；通过 Scala 和 Java 的配合完成 Akka 并发编程框架的实现，结合函数式编程和面向对象编程的思想理解 Akka 并发编程框架的应用，为后面学习 Spark 框架打好基础。

4.12 本章习题

1. 通过代码实现 Scala 作为值的函数、匿名函数、柯里化和闭包的开发。

2. 实践 Akka 并发编程框架。

第 5 章

Spark 安装部署与入门

Spark 是一种多语言引擎,用于在单节点机器或集群上执行数据工程、数据科学和机器学习,是一种用于大规模数据分析的统一引擎。

5.1 Spark 简介

Spark 是基于内存的迭代计算框架,适用于需要多次操作特定数据集的应用场合。需要反复操作的次数越多,所需读取的数据量越大,收益就越大。

5.1.1 MapReduce 与 Spark

Hadoop 中的 MapReduce 计算引擎存在如下问题。

第一,MapReduce 计算引擎的表达能力有限,仅仅支持 map 和 reduce 两种算子。而 Spark 计算引擎不仅支持 map 和 reduce,还支持 flatMap、filter、mapPartitions、distinct 等几十种算子。

第二,MapReduce 计算引擎的磁盘 I/O 开销大,因为其将中间计算结果写入磁盘,在需要的时候从磁盘读取,最后处理完还需要再次写入磁盘,这样频繁地访问磁盘会产生多次 I/O 开销。而 Spark 计算引擎在处理数据的过程中会把中间计算结果写入内存,在需要的时候从内存读取,最后处理完中间的计算任务一次性写入磁盘,因此节省了大量的 I/O 开销,其计算速度比 MapReduce 快了好几倍。

第三,MapReduce 计算引擎的延迟高,任务之间的衔接涉及 I/O 开销,在前一个任务执行完成之前,其他任务无法开始,难以胜任复杂、多阶段的计算任务。

Spark 吸取了 MapReduce 计算引擎的优点,同时也解决了 MapReduce 所面临的以上问题。Spark 的计算模式也属于 MapReduce,但不局限于 Map 和 Reduce 操作,还提供了多种数据集操作类型,编程模型比 MapReduce 更灵活。Spark 提供了内存计算,可将中间结果放到内存中,对于迭代运算效率更高。

Spark 基于 DAG(Directed Acyclic Graph,有向无环图)的任务调度执行机制,要优于 MapReduce 的迭代执行机制,如图 5-1 所示。

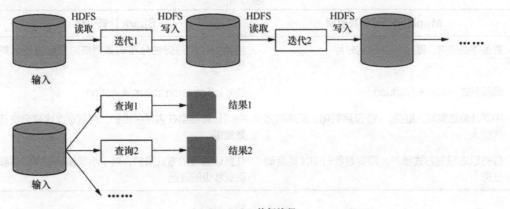

(a) MapReduce执行流程

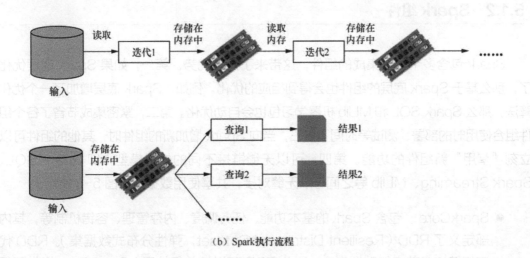

(b) Spark执行流程

图 5-1

如图 5-1 所示，若使用 MapReduce 进行迭代计算，非常耗资源，在迭代计算的过程中会频繁地读写磁盘。而 Spark 将数据载入内存中，在迭代计算的过程中可以直接使用内存中的中间计算结果进行运算，可避免从磁盘中频繁读取数据。在使用 MapReduce 与 Spark 执行逻辑回归算法的时间对比上出现了非常大的差异，如图 5-2 所示。

MapReduce 计算引擎与 Spark 计算引擎的对比如表 5-1 所示。

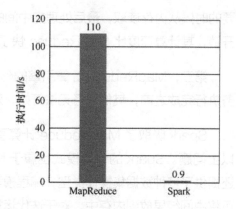

图 5-2　逻辑回归算法在 MapReduce 和 Spark 上执行的时间

表 5-1

MapReduce 计算引擎	Spark 计算引擎
数据存储结构：磁盘 HDFS 的分片	使用内存构建弹性分布式数据集对数据进行运算和缓存
编程范式：Map + Reduce	DAG: Transformation + Action
中间计算结果写入磁盘，I/O 及序列化、反序列化代价大	中间计算结果在内存中维护，存取速度比磁盘快几个数量级
任务以进程的方式维护，需要数秒时间才能启动任务	任务以线程的方式维护，对于小数据集的读取能够达到亚秒级的延迟

5.1.2　Spark 组件

Spark 包含多个紧密集成的组件，这带来了几个优势。第一，如果 Spark 底层优化了，那么基于 Spark 底层的组件也会得到相应的优化。例如，Spark 底层增加了一个优化算法，那么 Spark SQL 和 MLlib 机器学习包也会自动优化。第二，紧密集成节省了各个组件组合使用时的部署、测试等时间。第三，当向 Spark 增加新的组件时，其他的组件可以立刻"享用"新组件的功能。第四，可以无缝链接不同的处理模型，即 Spark SQL、Spark Streaming、MLlib 等之间可以无缝对接，共享使用数据，如图 5-3 所示。

- SparkCore：包含 Spark 的基本功能、任务调度、内存管理、容错机制等，其内部定义了 RDD（Resilient Distributed Dataset，弹性分布式数据集）。RDD 代表横跨许多节点的数据集合，因此 RDD 可以被并行处理。SparkCore 提供了很多 API 来创建和操作这些集合。

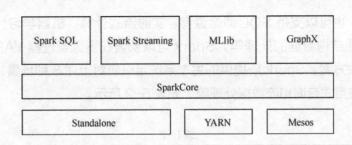

图 5-3

- Spark SQL：Spark 处理结构化数据的库，支持通过 SQL 查询数据，就像 HQL（Hive）一样，并且支持很多数据源，如 Hive 表、JSON 等。
- Spark Streaming：实时数据流处理组件，与 Storm 类似。Spark Streaming 提供了 API 来操作实时数据流数据。
- MLlib：通用机器学习功能包，其中包含分类、聚类、回归、协同过滤算法，以及一些机器学习原语（包括通用梯度下降优化算法）。MLlib 提供的这些算法都支持集群上的横向扩展。
- GraphX：处理图的库，例如社交网络图，并进行图的并行计算。GraphX 继承了 Spark RDD API，同时允许创建有向图。GraphX 提供了各种图的操作，如 subgraph 和 mapVertices，也包含常用的图算法，如 PangeRank 等。
- Standalone、YARN、Mesos：Spark 常见的 3 种运行模式，后面将重点介绍 Standalone 和 YARN。

5.1.3　Spark 生态系统

在实际应用中，大数据处理主要包括以下 3 种场景。

- 复杂的批量数据处理，通常时间跨度在数十分钟到数小时之间。
- 基于历史数据的交互式查询，通常时间跨度在数十秒到数分钟之间。
- 基于实时数据流的数据处理，通常时间跨度在数百毫秒到数秒之间。

当同时存在以上 3 种场景时，就需要同时部署 3 种不同的软件集群系统，比如 MapReduce、Impala 和 Storm。这样做难免会带来一些问题：不同场景之间数据 I/O 无法做到无缝共享，通常需要进行数据格式的转换；不同的软件需要不同的开发和维护团队，带来了较高的使用成本；难以对同一个集群中的各个系统进行统一的资源协调和分配。

为了解决以上问题，一个全新的集群框架出现了，它就是 Spark。Spark 的设计遵循"一个软件栈满足不同应用场景"的理念，逐渐形成了一套完整的生态系统。既能够提供内存计算框架，也可以支持 SQL 即席查询、实时流式计算、机器学习、基于历史数据的数据挖掘和图结构数据的处理等。Spark 可以部署在资源管理器 YARN 上，提供一站式大数据解决方案。Spark 所提供的生态系统足以应对上述 3 种场景，即同时支持批量数据处理、交互式查询和流数据处理等，如表 5-2 所示。

表 5-2

应用场景	时间跨度	其他框架	Spark 生态系统中的组件
复杂的批量数据处理	小时级	MapReduce、Hive	Spark
基于历史数据的交互式查询	分钟级、秒级	Impala、Dremel、Drill	Spark SQL
基于实时数据流的数据处理	毫秒级、秒级	Storm、S3	Spark Streaming
基于历史数据的数据挖掘	—	Mahout	MLlib
图结构数据的处理	—	Pregel、Hama	GraphX

Spark 生态系统已经成为 BDAS（Berkeley Data Analytics Stack，伯克利数据分析栈）的重要组成部分。Spark 生态系统主要包含 SparkCore、Spark SQL、Spark Streaming、MLlib 和 GraphX 组件，如图 5-4 所示。

访问接口层	Spark Streaming	BlinkDB	GraphX	MLbase
		Spark SQL		MLlib
引擎处理层	SparkCore			
数据存储层	Tachyon			
	HDFS、S3			
资源序列化层	Mesos		YARN	

图 5-4

5.1.4 Spark 架构

Spark 架构包括 Master（集群资源管理者）、Slave（运行任务的工作节点）、Driver Program（驱动控制程序）和 Executor（每个工作节点上负责任务的执行进程）。

Master 是集群资源管理者（Cluster Manager），Spark 支持 Standalone、YARN 和 Mesos。

Slave 在 Spark 中被称为 Worker，在其上开辟 Executor 负责任务（Task）的运行。

Driver Program 运行应用的 main 方法并且创建了 SparkContext。由 Cluster Manager 分配资源，SparkContext 将发送 Task 到 Executor 上执行，如图 5-5 所示。

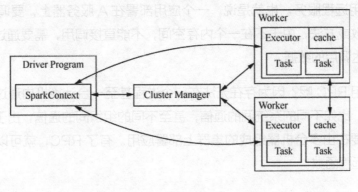

图 5-5

5.1.5　Spark 运行部署模式

Spark 运行部署模式有本地模式、Standalone 集群模式、Standalone-HA 集群模式、YARN 集群模式、Mesos 集群模式和 on cloud 集群模式。

本地模式主要用于开发和测试，分为 local 单线程和 local-cluster 多线程。

Standalone 集群模式，由 Spark 自己来管理自己的资源，通常用于开发和测试，使用典型的 Master-Slave 模型。

Standalone-HA 集群模式在实际生产环境中使用，其基于 Standalone 集群模式，使用 Zookeeper 搭建高可用集群，避免 Master 发生单点故障。

YARN 集群模式常用于生产环境，运行在 YARN 集群之上，由 YARN 负责资源管理，Spark 负责任务调度和计算。YARN 集群模式的好处是计算资源按需伸缩、集群利用率高、共享底层存储、避免数据跨集群迁移。

Mesos 集群模式，运行在 Mesos 资源管理器框架上，由 Mesos 负责资源管理，Spark 负责任务调度和计算，实际应用不多。

on cloud 集群模式，可以很方便地访问 Amazon S3 云存储，适合中小公司使用云服务的场景。

5.1.6 Spark 远程过程调用协议

RPC 是远程过程调用（Remote Procedure Call）的英文缩写，是指允许像调用本地服务一样调用远程服务。也就是说，一个应用部署在 A 服务器上，要调用 B 服务器上应用提供的函数或方法，由于不在一个内存空间，不能直接调用，需要通过网络来表达调用的语义和传达调用的数据。

为什么要用 RPC 呢？因为存在无法在一个进程甚至一台计算机内通过本地调用的方式完成的需求，比如不同的系统间的通信，甚至不同的组织间的通信，由于计算能力需要横向扩展，需要在由多台机器组成的集群上部署应用。有了 RPC，就可以像调用本地的函数一样调用远程函数。

5.2 Spark 环境搭建

Spark 支持多种部署模式。第一，是最简单的本地模式，即 Spark 所有进程都运行在一台机器的 JVM 中。第二，是伪分布式模式，即在一台机器中模拟集群运行，相关的进程运行在同一台机器上。第三，是分布式模式，包括 Spark 自带的 Standalone、YARN 和 Mesos。

5.2.1 本地模式部署

本地模式部署主要分为 4 个步骤。

1. 安装

首先登录 Spark 官方网站下载 Spark 安装包，选择 2.0 及以上版本。本书使用的是 Spark 3.0，下载的是 spark-3.0.3-bin-hadoop2.7.tgz。

解压安装包进行安装并重命名，操作命令如下所示。

```
cd /export/servers
tar spark-3.0.3-bin-hadoop2.7.tgz
mv spark-3.0.3-bin-hadoop2.7 spark
```

如果有权限问题，可以修改为 root，方便学习时操作，实际中使用运维分配的用户和权限即可。

```
chown -R root /export/servers/spark
chgrp -R root /export/servers/spark
```

解压之后我们会得到相关目录,具体如下。

bin: 可执行脚本。
conf: 配置文件。
data: 示例程序使用的数据。
examples: 示例程序。
jars: 依赖 jar 包。
python: Python API。
R: R 语言 API。
sbin: 集群管理命令。
yarn: 整合 YARN 需要的东西。

2. 启动 spark-shell

直接启动 bin 目录下的 spark-shell: ./spark-shell。直接使用 ./spark-shell,表示使用本地模式启动,在本机启动一个 Spark Submit 进程。--master 为指定参数。

```
spark-shell --master local[N]  #表示在本地模拟N个线程来运行当前任务
spark-shell --master local[*]  #表示使用当前机器上所有可用的资源
```

以下为不携带参数的情况。

```
spark-shell --master local[*]
```

还可以使用 --master 指定集群地址,表示把任务提交到集群上运行。

```
./spark-shell --master spark://node01:7077,node02:7077
```

3. 读取本地文本

准备一个本地文件,并对其进行读取。

```
hello spark
hello hadoop
hello hive
hello flink
hello
```

```
vim /root/words.txt
val textFile = sc.textFile("file:///root/words.txt")

    val counts = textFile.flatMap(_.split(" ")).map((_, 1)).reduceByKey(_ + _)
```

```
counts.collect //收集结果:Array[(String, Int)] = Array((spark,1), (hello,5),
(hive,1), (hadoop,1), (flink,1))
```

4. 读取 HDFS 上的文件

上传文件到 HDFS（Hadoop Distributed File System，Hadoop 分布式文件系统）。

```
hadoop fs -put /root/words.txt /wordcount/input/words.txt
```

如果目录不存在，可以自行创建。

```
hadoop fs -mkdir -p /wordcount/input
```

读取文件。

```
val textFile = sc.textFile("hdfs://node01:8020/wordcount/input/words.txt")
    val counts = textFile.flatMap(_.split(" ")).map((_, 1)).reduceByKey(_ + _)
    counts.collect().foreach(print _)
```

结束后可以删除测试文件夹。

```
hadoop fs -rm -r /wordcount
```

5.2.2 Standalone 集群模式

Standalone 集群模式的详解如下。

1. 集群角色介绍

Spark 是基于内存计算的大数据并行计算框架，实际中运行计算任务会使用集群模式，那么我们先来学习 Spark 自带的 Standalone 集群模式，了解一下它的架构及运行机制，如图 5-6 所示。

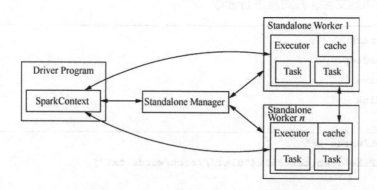

图 5-6

Standalone 集群使用了分布式计算中的 Master-slave 模型。Master 是集群中含有 Master 进程的节点。Slave 是集群中的 Worker 节点，含有 Executor 进程。

2. 集群规划

利用虚拟化技术构建 3 个节点，其中 node01 为主节点，node02 和 node03 为从节点，如下。

```
node01:master
node02:slave/worker
node03:slave/worker
```

3. 修改配置文件并分发

修改 Spark 配置文件的方法如下。

```
cd /export/servers/spark/conf
mv spark-env.sh.template spark-env.sh
vim    spark-env.sh
#配置Java环境变量
export JAVA_HOME=/export/servers/jdk1.8
#指定Spark Master的IP地址
export SPARK_MASTER_HOST=node01
#指定Spark Master的端口
export SPARK_MASTER_PORT=7077
mv slaves.template slaves
vim    slaves
node02
node03
```

通过 scp 命令，可将配置文件分发到其他机器上。

```
scp -r /export/servers/spark node02:/export/servers
scp -r /export/servers/spark node03:/export/servers
scp /etc/profile root@node02:/etc
scp /etc/profile root@node03:/etc
source /etc/profile #刷新配置
```

4. 启动和停止

在 Master 节点上启动 Spark 集群的方法如下。

```
/export/servers/spark/sbin/start-all.sh
```

在 Master 节点上停止 Spark 集群的方法如下。

```
/export/servers/spark/sbin/stop-all.sh
```

在 Master 节点上启动和停止 Master 的方法如下。

```
start-master.sh
stop-master.sh
```

在 Slave 节点上启动和停止 Worker。

```
start-slaves.sh
stop-slaves.sh
```

5. 查看 Web 界面

正常启动 Spark 集群后，通过 Spark 的 Web 界面（http://node01:8080/）可查看相关信息。

6. 测试

例如，使用集群模式运行 Spark 程序，读取 HDFS 上的文件并进行单词统计。

使用集群模式启动 spark-shell 的命令如下。

```
/export/servers/spark/bin/spark-shell --master spark://node01:7077
```

读取 HDFS 上的 wordcount 文件下的文件内容进行处理，代码如下。

```
sc.textFile("hdfs://node01:8020/wordcount/input/words.txt")
.flatMap(_.split(" ")).map((_, 1))
.reduceByKey(_ + _).saveAsTextFile("hdfs://node01:8020/wordcount/results")
```

注意：在集群模式下不要直接读取本地文件，应该读取 HDFS 上的文件，因为程序运行在集群上，具体在哪个节点上运行我们并不知道，其他节点可能并没有那个要读取的数据文件。

5.2.3 Standalone-HA 集群模式

Standalone-HA 集群模式一般用于生产环境中的开发和测试。

1. 原理

Spark Standalone 集群是 Master-Slave 架构的集群模式，和大部分的 Master-Slave

架构的集群一样，存在 Master 单点故障的问题。

如何解决这个单点故障的问题呢？Spark 提供了两种方案。

一是基于文件系统的单点恢复，该方案只能用于开发或测试环境。

二是基于 ZooKeeper 的 Standby Masters，该方案可以用于生产环境。

基于 ZooKeeper 的 Standby Masters 的原理如图 5-7 所示。

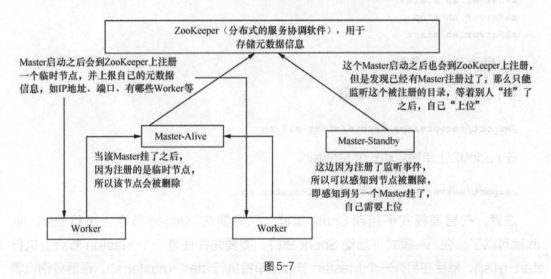

图 5-7

2. 配置 HA

该 HA（High Availability，高可用）方案使用起来很简单，首先启动一个 ZooKeeper 集群，然后在不同节点上启动 Master，注意这些节点需要具有相同的 ZooKeeper 配置。

先停止 Spark 集群。

`/export/servers/spark/sbin/stop-all.sh`

在 node01 上进行配置。

`vim /export/servers/spark/conf/spark-env.sh`

注释掉 Master 配置。

`#export SPARK_MASTER_HOST=node01`

在 spark-env.sh 中添加 SPARK_DAEMON_JAVA_OPTS，内容如下。

```
export SPARK_DAEMON_JAVA_OPTS="-D spark.deploy.recoveryMode=ZOOKEEPER
-D spark.deploy.zookeeper.url=node01:2181,node02:2181,node03:2181
-D spark.deploy.zookeeper.dir=/spark"
```

其中，spark.deploy.recoveryMode 表示恢复模式；spark.deploy.zookeeper.url 是 ZooKeeper 的 Server 地址；spark.deploy.zookeeper.dir 用于保存集群元数据信息包括 Worker、Driver、Application 信息。

3. 启动 ZooKeeper 集群

启动 ZooKeeper 集群的方法如下。

```
zkServer.sh status
zkServer.sh stop
zkServer.sh start
```

4. 启动 Spark 集群

在 node01 上启动 Spark 集群。

```
/export/servers/spark/sbin/start-all.sh
```

在 node02 上单独启动一个 Master。

```
/export/servers/spark/sbin/start-master.sh
```

注意，在普通模式下启动 Spark 集群，只需要在 Masten 节点上执行 start-all.sh 就可以了；在 HA 模式下启动 Spark 集群，需要先在任意一个 Masten 节点上执行 start-all.sh，然后在另外一个 Masten 节点上单独执行 start-master.sh。最后分别查看 node01 和 node02 的状态，可以观察到有一个节点的状态为 Standby。

5. 测试 HA

① 在 node01 上使用 jps 命令查看 Master 进程 ID。

② 使用 kill -9 ID 强制结束该进程。

③ 稍等片刻后刷新 node02 的 Web 界面，发现 node02 处于 Alive 状态。

④ 测试集群模式，提交任务。集群模式下启动 spark-shell。运行之前的脚本，存储路径不能和之前的相同，必须为新路径。

5.2.4　YARN 集群模式

YARN 集群模式是生产环境中常用的一种模式。

1. 准备工作

首先安装并启动 Hadoop（需要使用 HDFS 和 YARN），然后安装单机版 Spark（预先安装）。

注意：不需要集群，因为把 Spark 程序提交给 YARN 运行本质上是把字节码提交给 YARN 集群上的 JVM 运行，但是需要有一个东西帮我们把任务提交给 YARN，所以需要一个单机版 Spark，里面有 spark-shell 命令、spark-submit 命令等。

在 spark-env.sh 中添加 HADOOP_CONF_DIR 配置，操作命令为：vim /export/servers/spark/conf/spark-env.sh。

在 spark-env.sh 中配置 hadoop 的文件位置，代码如下。

```
export HADOOP_CONF_DIR=/export/servers/hadoop/etc/hadoop
```

2. cluster 模式

在企业生产环境中大部分都是使用 cluster 模式运行 Spark 应用。Spark YARN 的 cluster 模式指的是 Driver 程序运行在 YARN 集群上，如图 5-8 所示。

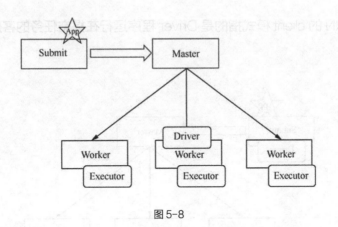

图 5-8

之前我们使用的 spark-shell 是一个简单的用来测试的交互式窗口，正式生产情况下需要提交 jar 包来运行。

```
export/servers/spark/bin/spark-submit \        # Spark环境Submit命令
--class org.apache.spark.examples.SparkPi \    # 类入口
--master yarn \                                # Spark YARN
--deploy-mode cluster \                        # 使用cluster 模式
--driver-memory 1g \                           # Driver 的内存
--executor-memory 1g \                         # 每个Executor 的内存
--executor-cores 2 \                           # 一共使用多少个Executor
```

```
--queue default \  # Spark YARN的队列
/export/servers/spark/examples/jars/spark-examples_2.11-2.2.0.jar  # 执行的jar包
```

查看 Web 界面（http://node-01:8088/cluster），如图 5-9 所示。

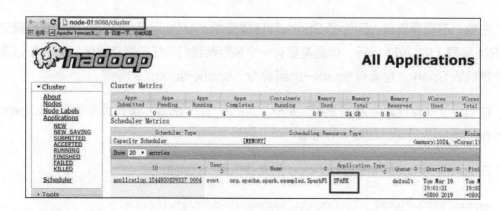

图 5-9

3. client 模式

在调试阶段，想要更方便地观察日志情况，也可以使用 client 模式。

Spark YARN 的 client 模式指的是 Driver 程序运行在提交任务的客户端，如图 5-10 所示。

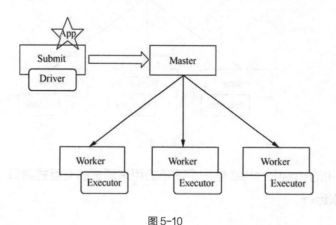

图 5-10

```
export/servers/spark/bin/spark-submit \  # Spark环境Submit命令
--class org.apache.spark.examples.SparkPi \  # 类入口
--master yarn \  # Spark YARN
--deploy-mode client\ # 使用client模式
--driver-memory 1g \  # Driver 的内存
--executor-memory 1g \ # 每个Executor 的内存
```

```
--executor-cores 2 \ # 一共使用多少个Executor
--queue default \ # Spark YARN 的队列
```

```
/export/servers/spark/examples/jars/spark-examples_2.11-2.2.0.jar  # 执行的jar 包
```

想要变更为 client 模式，只需要将 --deploy-mode 设置成 client 即可。同样，提交后，也可以在 Spark UI 上看到对应的任务。

4. 两种模式的区别

cluster 和 client 模式最本质的区别在于 Driver 程序运行的位置！ 运行在 YARN 集群中的就是 cluster 模式，运行在客户端的就是 client 模式。

生产环境中使用 cluster 模式。Driver 程序在 YARN 集群中应用的运行结果不能在客户端显示，该模式下 Driver 运行 ApplicationMaster 进程。如果出现问题，YARN 会重启 ApplicationMaster（Driver）。

Driver 运行在 client 模式的 Spark Submit 进程中，应用的运行结果会在客户端显示。更为重要的是，要知道这两种模式的日志打印的位置，cluster 模式的日志需要我们进入 Spark UI 后查看，而 client 模式的日志可以在提交的机器上查看。

5. 错误解决

如果整合报错或无法查看日志，需进行如下操作。首先，修改 Hadoop 的 yarn-site.xml。

```
vim /export/servers/hadoop/etc/hadoop/yarn-site.xml
<property>
<name>yarn.resourcemanager.hostname</name>
<value>node01</value>
</property>
<property>
<name>yarn.nodemanager.aux-services</name>
<value>mapreduce_shuffle</value>
</property>
<!-- 关闭YARN内存检查 -->
<property>
<name>yarn.nodemanager.pmem-check-enabled</name>
<value>false</value>
</property>
<property>
<name>yarn.nodemanager.vmem-check-enabled</name>
```

```xml
<value>false</value>
</property>
<!-- 如果开启如下配置则需要开启Spark历史服务器-->
<property>
<name>yarn.log-aggregation-enable</name>
<value>true</value>
</property>
<property>
<name>yarn.log-aggregation.retain-seconds</name>
<value>604800</value>
</property>
<property>
<name>yarn.log.server.url</name>
<value>http://node01:19888/jobhistory/logs</value>
</property>
```

接着，分发并重启 Hadoop 服务。

```
/export/servers/hadoop/sbin/stop-dfs.sh
/export/servers/hadoop/sbin/stop-yarn.sh
/export/servers/hadoop/sbin/start-dfs.sh
/export/servers/hadoop/sbin/start-yarn.sh
```

如果要整合 YARN 历史服务器和 Spark 历史服务器，则还需要进行如下操作。

配置 YARN 历史服务器并启动。

```
/export/servers/hadoop/sbin/mr-jobhistory-daemon.sh start historyserver
```

配置 Spark 历史服务器，修改 spark-defaults.conf。

```
vim /export/servers/spark/conf/spark-defaults.conf
spark.yarn.historyServer.address node01:4000
```

启动 Spark 历史服务器。

```
/export/servers/spark/sbin/start-history-server.sh
```

如果依赖的 jar 包较多，可以上传到 HDFS 并告诉 YARN 去获取 spark-defaults.conf 中的配置。

```
spark.yarn.jars = hdfs://node01:8020/sparkjars/*
```

配置之后，各个节点会去 HDFS 下载并缓存配置文件，如果不配置，Spark 程序启动后会把 Spark_HOME 打包并分发到各个节点。

5.2.5 Spark 命令

1. spark-shell

之前我们使用 spark-shell 提交任务。spark-shell 是 Spark 自带的交互式 Shell 命令，方便用户进行交互式编程，我们可以通过该命令进入 Spark-Shell 交互式编程窗口，就可以用 Scala 语言编写 Spark 应用程序，该命令适合测试时使用。

```
spark-shell #可以携带参数
spark-shell --master local[N] #数字N表示在本地模拟N个线程来运行当前任务
spark-shell --master local[*] #*表示使用当前机器上所有可用的资源，默认不携带参数就
是--master local[*]
spark-shell --master spark://node01:7077,node02:7077 #表示运行在集群上
```

2. spark-submit

spark-shell 交互式编程确实很方便我们测试，但是在实际中一般使用 IDEA 开发 Spark 应用程序，并将其打包成 jar 包交给 Spark 集群 /YARN 集群去执行。

3. master

master 参数形式如表 5-3 所示。

表 5-3

master 参数形式	解释
local	本地以一个 Worker 线程运行 (例如非并行的情况)
local[N]	本地以 N 个 Worker 线程运行 (理想情况下，N 设置为机器的 CPU 核数)
local[*]	本地以本机同样核数的线程运行
spark://HOST:PORT	连接到指定的 Spark Standalone 集群。端口是集群配置的端口，默认值为 7077
mesos://HOST:PORT	连接到指定的 Mesos 集群。端口是用户配置的 Mesos 端口，默认值为 5050，或者使用 ZooKeeper，格式为 mesos://zk://...
yarn-client	以 client 模式连接到 YARN 集群，集群的位置基于 HADOOP_CONF_DIR 变量找到
yarn-cluster	以 cluster 模式连接到 YARN 集群，集群的位置基于 HADOOP_CONF_DIR 变量找到

其他参数如表 5-4 所示。

表 5-4

参数	解释
--master	表示要连接的集群管理器
--class	运行 Java 或 Scala 程序时应用的主类
--name	应用的显示名，会显示在 Spark 的网页用户界面中
--jars	需要上传并放到应用的 CLASSPATH 中的 jar 包的列表。如果应用依赖于少量第三方的 jar 包，可以把它们放在这个参数里
--files	需要放到应用工作目录中的文件的列表。这个参数一般用来放需要分发到各节点的数据文件
--py-files	需要添加到 PYTHONPATH 中的文件的列表。其中可以包含 .py、.egg 以及 .zip 文件
--executor-memory	Executor 进程使用的内存量，以字节为单位。可以使用后缀指定更大的单位，比如 "512m"（512MB）、"15g"（15GB）等
--driver-memory	Driver 进程使用的内存量，以字节为单位。可以使用后缀指定更大的单位，比如 "512m"（512MB）、"15g"（15GB）等

4. 配置历史服务器

Spark 程序运行完毕且窗口关闭后，我们就无法再查看运行记录的 Web UI（4040）。但是历史服务器可以提供一个服务，即通过读取日志文件，使得我们在程序运行结束后依然能够查看运行过程。

① 修改文件名。

```
cd /export/servers/spark/conf
cp spark-defaults.conf.template spark-defaults.conf
```

② 修改配置。

```
vim spark-defaults.conf
spark.eventLog.enabled true
spark.eventLog.dir hdfs://node01:8020/sparklog
```

③ HDFS 上的目录需要先手动创建。

```
hadoop fs -mkdir -p /sparklog
```

④ 修改 spark-env.sh。

```
vim spark-env.sh
export SPARK_HISTORY_OPTS="-Dspark.history.ui.port=4000
-Dspark.history.retainedApplications=3
-Dspark.history.fs.logDirectory=hdfs://node01:8020/sparklog
```

⑤ 同步文件。

```
scp -r /export/servers/spark/conf/
@node02:/export/servers/spark/conf/
scp -r /export/servers/spark/conf/
@node03:/export/servers/spark/conf/
```

⑥ 重启集群。

```
/export/servers/spark/sbin/stop-all.sh
/export/servers/spark/sbin/start-all.sh
```

⑦ 在 Master 上启动历史服务器。

```
/export/servers/spark/sbin/start-history-server.sh
```

⑧ 在 4000 端口查看历史日志（如果加载不出来则换浏览器试试）。

```
http://node01:4000/
```

⑨ 你可能会遇到 Hadoop HDFS 的写入权限问题。

```
org.apache.hadoop.security.AccessControlException
```

⑩ 在 hdfs-site.xml 中添加如下配置，关闭权限验证。生产环境下不建议如此，请确认 Hadoop 的具体信息。

```
<property>
<name>dfs.permissions</name>
<value>false</value>
</property>
```

5.3 编写 Spark 应用程序

Spark 同时支持 Scala、Python、Java 这 3 门语言。本节将采用 Scala 语言来编写 Spark 应用程序。Spark 对 Scala 语言的支持是最好的，因为 Scala 有丰富和易用的编程接口。

5.3.1 Maven 简介

Maven 是一个跨平台的项目管理工具，主要服务于基于 Java 平台的项目构建、依赖管理和项目信息管理。无论是在小型的开源类库项目中，还是在大型的企业级应用中，Maven 都能大显身手。

很多 Java 应用都会借助第三方的开源类库，这些类库都可通过依赖的方式引入项目中。随着依赖的增多，版本不一致、版本冲突、依赖臃肿等问题接踵而来。

手动解决这些问题是十分枯燥和烦琐的。Maven 提供了一种优秀的解决方案，通过一组坐标，Maven 能够找到任何一个 Java 类库。

使用 Maven 还能享受一个额外的好处，即 Maven 对于项目目录结构、测试用例命名方式等内容都有既定的规则，只要遵循这些规则，在项目间切换的时候就免去了额外的学习成本。

5.3.2 安装 Maven

安装 Maven 的步骤如下。

① 去 Maven 官网下载并安装 Maven。

② 设置环境变量，并使环境变量生效。

③ 执行 mvn -v 命令。

注意：仅在安装 IDEA 的节点上安装 Maven。

Maven 会根据项目的 pom.xml 文件下载很多的 jar 包，下载后的 jar 包放在 ~/.m2/repository 中。

将 $MAVEN_HOME\conf\settings.xml 文件复制到 ~/.m2 目录。修改默认下载地址，打开 ~/.m2/settings.xml，在文件的第 16 行，进行如下修改。

```
<mirrors>
  <mirror>
    <id>alimaven</id>
```

```
      <name>aliyun maven</name>
      <url>http://maven.aliyun.com/nexus/content/groups/public/</url>
      <mirrorOf>central</mirrorOf>
   </mirror>
</mirrors>
```

5.3.3 Spark 开发环境搭建

前面章节已经介绍了 Scala 插件的安装方法。接下来，我们进行全局 JDK 和 Library 的配置。为了避免每次重复配置 JDK，这里进行一次全局配置。在 IDEA 欢迎界面的右下角单击 Configure，然后在 Project Defaults 的子菜单中选择 Project Structure，如图 5-11 所示。

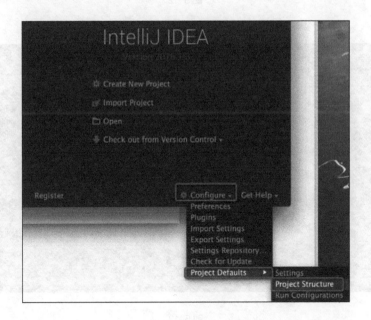

图 5-11

在打开的 Default Project Structure 界面的左侧选择 Project，在右侧创建一个新的 JDK（本机已经安装过 JDK），如图 5-12 所示。在菜单中单击 JDK 后，在打开的对话框中选择安装 JDK 的位置。注意，JDK 安装的根目录就是 JAVA_HOME 中设置的目录。

配置全局的 Scala SDK。在 IDEA 欢迎页面的右下角单击 Configure，在 Project Defaults 的子菜单中选择 Project Structure，在打开的界面的左侧选择 Global Libraries，中间一栏中有一个加号标志，单击后在菜单中选择 Scala SDK，在打开的对话框中选择所安装的 Scala，单击 OK 按钮，这时候会在中间一栏位置处出现 Scala 的

SDK，在其上右击后选择 Copy to Project Libraries，这个操作是为了将 Scala SDK 添加到项目的默认 Library，如图 5-13 所示。

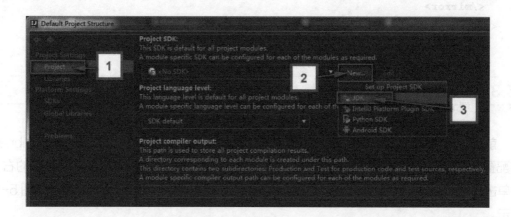

图 5-12

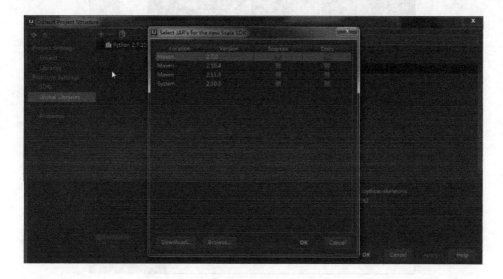

图 5-13

新建 Maven 项目。在 IDEA 欢迎界面单击 Create New Project，在打开的界面的左侧选择 Maven，然后在右侧的 Project SDK 一项中，查看是否有正确的 JDK 配置项，如图 5-14 所示。正常来说，这一栏会自动填充，因为在前面已经配置过全局的 Project JDK。如果这里没有正常显示 JDK，可以单击右侧的 New 按钮，然后指定 JDK 安装路径的根目录，并单击 Next 按钮，来到 Maven 项目最重要的 3 个参数的设置页面，3 个参数分别为 GroupId、ArtifactId 和 Version。

图 5-14

5.3.4 配置 pom.xml 文件

创建 Maven 项目并补全目录、配置 pom.xml 文件。

```xml
<?xml version="1.0" encoding="UTF-8"?>
<project
    xmlns="http://maven.apache.org/POM/4.0.0"
    xmlns:xsi="http://www.w3.org/2001/XMLSchema-instance" xsi:schemaLocation="http://maven.apache.org/POM/4.0.0
    http://maven.apache.org/xsd/maven-4.0.0.xsd">
    <modelVersion>4.0.0</modelVersion>
    <groupId>cn.itcast</groupId>
    <artifactId>SparkDemo</artifactId>
    <version>1.0-SNAPSHOT</version>
    <!-- 指定仓库位置，依次为aliyun、cloudera和jboss仓库 -->
    <repositories>
        <repository>
            <id>aliyun</id>
            <url>http://maven.aliyun.com/nexus/content/groups/public/</url>
        </repository>
        <repository>
            <id>cloudera</id>
            <url>https://repository.cloudera.com/artifactory/cloudera-repos/</url>
        </repository>
        <repository>
            <id>jboss</id>
```

```xml
        <url>http://repository.jboss.com/nexus/content/groups/public</url>
    </repository>
</repositories>
<properties>
    <maven.compiler.source>1.8</maven.compiler.source>
    <maven.compiler.target>1.8</maven.compiler.target>
    <encoding>UTF-8</encoding>
    <scala.version>2.11.8</scala.version>
    <scala.compat.version>2.11</scala.compat.version>
    <hadoop.version>2.7.4</hadoop.version>
    <spark.version>2.2.0</spark.version>
</properties>
<dependencies>
    <dependency>
        <groupId>org.scala-lang</groupId>
        <artifactId>scala-library</artifactId>
        <version>${scala.version}</version>
    </dependency>
    <dependency>
        <groupId>org.apache.spark</groupId>
        <artifactId>spark-core_2.11</artifactId>
        <version>${spark.version}</version>
    </dependency>
    <dependency>
        <groupId>org.apache.spark</groupId>
        <artifactId>spark-sql_2.11</artifactId>
        <version>${spark.version}</version>
    </dependency>
    <dependency>
        <groupId>org.apache.spark</groupId>
        <artifactId>spark-hive_2.11</artifactId>
        <version>${spark.version}</version>
    </dependency>
    <dependency>
        <groupId>org.apache.spark</groupId>
        <artifactId>spark-hive-thriftserver_2.11</artifactId>
        <version>${spark.version}</version>
    </dependency>
    <dependency>
        <groupId>org.apache.spark</groupId>
        <artifactId>spark-streaming_2.11</artifactId>
```

```xml
            <version>${spark.version}</version>
        </dependency>
        <!-- <dependency><groupId>org.apache.spark</groupId><artifactId>spark-streaming-kafka-0-8_2.11</artifactId><version>${spark.version}</version></dependency>-->
        <dependency>
            <groupId>org.apache.spark</groupId>
            <artifactId>spark-streaming-kafka-0-10_2.11</artifactId>
            <version>${spark.version}</version>
        </dependency>
        <dependency>
            <groupId>org.apache.spark</groupId>
            <artifactId>spark-sql-kafka-0-10_2.11</artifactId>
            <version>${spark.version}</version>
        </dependency>
        <dependency>
            <groupId>org.apache.hadoop</groupId>
            <artifactId>hadoop-client</artifactId>
            <version>2.7.4</version>
        </dependency>
        <dependency>
            <groupId>org.apache.hbase</groupId>
            <artifactId>hbase-client</artifactId>
            <version>1.3.1</version>
        </dependency>
        <dependency>
            <groupId>org.apache.hbase</groupId>
            <artifactId>hbase-server</artifactId>
            <version>1.3.1</version>
        </dependency>
        <dependency>
            <groupId>com.typesafe</groupId>
            <artifactId>config</artifactId>
            <version>1.3.3</version>
        </dependency>
        <dependency>
            <groupId>mysql</groupId>
            <artifactId>mysql-connector-java</artifactId>
            <version>5.1.38</version>
        </dependency>
    </dependencies>
    <build>
```

```xml
<sourceDirectory>src/main/scala</sourceDirectory>
<testSourceDirectory>src/test/scala</testSourceDirectory>
<plugins>
    <!-- 指定编译Java的插件 -->
    <plugin>
        <groupId>org.apache.maven.plugins</groupId>
        <artifactId>maven-compiler-plugin</artifactId>
        <version>3.5.1</version>
    </plugin>
    <!-- 指定编译Scala的插件 -->
    <plugin>
        <groupId>net.alchim31.maven</groupId>
        <artifactId>scala-maven-plugin</artifactId>
        <version>3.2.2</version>
        <executions>
            <execution>
                <goals>
                    <goal>compile</goal>
                    <goal>testCompile</goal>
                </goals>
                <configuration>
                    <args>
                        <arg>-dependencyfile</arg>
                        <arg>${project.build.directory}/.scala_dependencies</arg>
                    </args>
                </configuration>
            </execution>
        </executions>
    </plugin>
    <plugin>
        <groupId>org.apache.maven.plugins</groupId>
        <artifactId>maven-surefire-plugin</artifactId>
        <version>2.18.1</version>
        <configuration>
            <useFile>false</useFile>
            <disableXmlReport>true</disableXmlReport>
            <includes>
                <include>**/*Test.*</include>
                <include>**/*Suite.*</include>
            </includes>
```

```xml
                </configuration>
            </plugin>
            <plugin>
                <groupId>org.apache.maven.plugins</groupId>
                <artifactId>maven-shade-plugin</artifactId>
                <version>2.3</version>
                <executions>
                    <execution>
                        <phase>package</phase>
                        <goals>
                            <goal>shade</goal>
                        </goals>
                        <configuration>
                            <filters>
                                <filter>
                                    <artifact>*:*</artifact>
                                    <excludes>
                                        <exclude>META-INF/*.SF</exclude>
                                        <exclude>META-INF/*.DSA</exclude>
                                        <exclude>META-INF/*.RSA</exclude>
                                    </excludes>
                                </filter>
                            </filters>
                        </configuration>
                    </execution>
                </executions>
            </plugin>
        </plugins>
    </build>
</project>
```

其核心是 spark-core_2.11，其余的可以根据需求进行添加。

5.3.5 开发应用程序——本地运行

将程序运行结果输出到控制台。

```
import org.apache.spark.rdd.RDD
import org.apache.spark.{SparkConf, SparkContext}
object WordCount {
```

```scala
def main(args: Array[String]): Unit = {
//1.创建SparkContext
val config = new SparkConf().setAppName("wc").setMaster("local[*]")
val sc = new SparkContext(config) sc.setLogLevel("WARN")
//2.读取文件
//RDD可以简单理解为分布式的集合,Spark对它做了很多的封装,
//让程序员使用起来就像操作本地集合一样简单
val fileRDD: RDD[String] = sc.textFile("./words.txt")
//3.处理数据
//对每一行按空格进行切分并压平,以形成一个新的集合,其中存放一个个单词
//flatMap是对集合中的每一个元素进行操作,再进行压平
val wordRDD: RDD[String] = fileRDD.flatMap(_.split(" "))
//每个单词记为1
val wordAndOneRDD: RDD[(String, Int)] = wordRDD.map((_,1))
//根据key进行聚合,统计每个单词的数量
//wordAndOneRDD.reduceByKey((a,b)=>a+b)
//第一个_:之前累加的结果
//第二个_:当前进来的数据
val wordAndCount: RDD[(String, Int)] = wordAndOneRDD.reduceByKey(_+_)
//4.收集结果
val result: Array[(String, Int)] =
wordAndCount.collect() result.foreach(println)
}
}
```

5.3.6 修改应用程序——集群运行

修改以上代码,将结果输出到目标文件系统,如 HDFS。

```scala
import org.apache.spark.rdd.RDD
import org.apache.spark.{SparkConf, SparkContext}
object WordCount {
def main(args: Array[String]): Unit = {
//1.创建SparkContext
val config = new SparkConf().setAppName("wc")//.setMaster("local[*]")
val sc = new SparkContext(config) sc.setLogLevel("WARN")
//2.读取文件
//RDD可以简单理解为分布式的集合,Spark对它做了很多的封装,
```

```
//让程序员使用起来就像操作本地集合一样简单
val fileRDD: RDD[String] = sc.textFile(args(0)) //文件输入路径
//3.处理数据
//对每一行按空格进行切分并压平，以形成一个新的集合，其中存放一个个单词
//flatMap是对集合中的每一个元素进行操作，再进行压平
val wordRDD: RDD[String] = fileRDD.flatMap(_.split(" "))
//每个单词记为1
val wordAndOneRDD: RDD[(String, Int)] = wordRDD.map((_,1))
//根据key进行聚合，统计每个单词的数量
//wordAndOneRDD.reduceByKey((a,b)=>a+b)
//第一个_:之前累加的结果
//第二个_:当前进来的数据
val wordAndCount: RDD[(String, Int)] = wordAndOneRDD.reduceByKey(_+_)
wordAndCount.saveAsTextFile(args(1))//文件输出路径
//4.收集结果
//val result: Array[(String, Int)] = wordAndCount.collect()
//result.foreach(println)
  }
}
```

在 IDEA 开发工具中对程序代码进行打包，如图 5-15 所示。

将打包好的 jar 包上传，并提交到 Spark 集群，代码如下。

```
/export/servers/spark/bin/spark-submit \
--class WordCount \
--master spark://node01:7077,node02:7077 \
--executor-memory 1g \
--total-executor-cores 2 \
/root/wc.jar \
hdfs://node01:8020/wordcount/input/words.txt   \ hdfs://node01:8020/wordcount/output4
```

图 5-15

提交到 YARN 集群，代码如下所示。

```
/export/servers/spark/bin/spark-submit \
--class WordCount \
--master yarn \
```

```
--deploy-mode cluster \
--driver-memory 1g \
--executor-memory 1g \
--executor-cores 2 \
--queue default \
/root/wc.jar \
hdfs://node01:8020/wordcount/input/words.txt    \ hdfs://node01:8020/wordcount/output5
```

5.3.7 集群硬件配置说明

1. 存储

在大数据领域,有一句名言:"移动数据不如移动计算。"

如果将数据从一个节点移动到另外一个节点,甚至从一个局域网移动到另外一个局域网,必然会牵涉大量的磁盘 I/O 和网络 I/O,非常影响性能。而这里的计算,可以理解为封装了业务代码的 jar 包,这是轻量级的,可有效缓解 I/O 带来的弊端。因此,将 Spark 集群节点尽可能部署到靠近存储系统的节点是非常重要的,因为大部分数据通常从外部存储系统获取。

以 HDFS 作为存储系统为例,建议在与 HDFS 相同的节点上运行 Spark。最简单的方式就是将 Spark 的 Standalone 集群和 Hadoop 集群部署在相同节点上,同时配置好 Spark 和 Hadoop 的内存及 CPU,以避免相互干扰。

此外,也可以将 Spark 和 Hadoop 运行在共同的集群资源管理器上,如 YARN 和 Mesos。

对于低延迟数据存储,如 HBase,优先在与存储系统不同的节点上运行计算任务,以避免干扰(计算引擎在处理任务时,比较消耗服务器资源,可能影响低延迟存储系统的即时响应)。

2. 本地磁盘

尽管 Spark 可以在内存中处理大量的计算,但它仍然需要使用本地磁盘来存储不适合缓存的数据。建议每个节点配备 4 ~ 8 块磁盘,并且将这些磁盘作为独立的磁盘挂载在节点上即可,不需要做磁盘阵列。

在 Linux 中，使用 noatime 选项安装磁盘，以减少不必要的写操作。在 Spark 中，通过参数 spark.local.dir 可以配置多个本地磁盘目录，多个目录之间以逗号分隔。如果 Spark 任务运行在 HDFS 上，与 HDFS 保持一致即可。

使用 noatime 选项安装磁盘，要求当挂载文件系统时，可以指定标准 Linux 安装选项，这将停止该文件系统上的 atime 更新。

磁盘挂载命令为 mount -t gfs BlockDevice MountPoint -o noatime（BlockDevice 用于指定 GFS 驻留的块设备，MountPoint 用于指定 GFS 应安装的目录）。

3. 内存

通常情况下，每台机器的内存配置从 8GB 到数百 GB，Spark 都能良好地运行。但建议最多分配给 Spark 75% 的内存，剩余的留给操作系统和缓冲存储。

当然，具体需要多少内存取决于你的应用程序。要确定应用程序使用的特定数据集需要多大内存，请加载部分数据集到内存缓存起来，然后在 Spark UI（http://driver-node:4040）的 Storage 界面去看它的内存占用量。

注意，内存使用多少受到存储级别和序列化格式的影响。

对于超过 200GB 的内存，JVM 运行状态并不是一直表现良好。如果你的机器内存超过了 200GB，那么可以在一个节点上运行多个 Worker。在 Spark Standalone 模式下，可以在配置文件 conf/spark-env.sh 中设置 SPARK_WORKER_INSTANCES 的值来设置每个节点 Worker 的数目，通过 SPARK_WORKER_CORES 参数来设置每个 Worker 的核数。

4. 网络

根据以往的经验，如果数据存储在内存中，那么 Spark 应用程序的瓶颈往往就在网络。特别是针对 Reduce 操作，如 group-by、reduce-by 和 SQL join，就更加明显。在任何给定的应用程序中，都可以通过 Spark UI 查看 Spark shuffle 过程中跨网络传输了多少数据。

5. CPU 核

因为 Spark 在线程之间执行最小的共享 CPU，所以它可以扩展到每台机器几十个 CPU 核。建议每台机器至少配置 8 个 CPU 核。当然，具体根据任务的 CPU 负载来配置。一旦数据存储在内存中，大多数应用程序的瓶颈就在 CPU 和网络。

5.4 本章总结

本章重点介绍了 Spark 的基本概念，包含 Spark 的特性、架构、生态系统、运行模式等内容；其次介绍了 Spark 的环境搭建，涉及本地模式、Spark 自带的 Standalone 集群模式和 YARN 集群模式；最后介绍了如何在 IDEA 开发工具中开发 Spark 应用程序，将其打包并上传到集群中运行。

5.5 本章习题

1. 复述 Spark 的特性、组件、架构以及运行模式。

2. 在本地机器上分别采用本地模式、Standalone 集群以及 YARN 集群模式部署 Spark 环境。

3. 在本地完成 WordCount 案例，并采用本地模式和集群模式进行测试。

第 6 章 SparkCore 编程

海量数据处理在许多迭代式算法（比如机器学习、图算法等）和交互式数据挖掘中，不同计算阶段之间会重用中间结果，即一个阶段的输出会作为下一个阶段的输入。但是，MapReduce 计算框架采用非循环式的数据流模型，把中间结果写入 HDFS，带来了大量的数据复制、磁盘 I/O 和序列化开销。且 MapReduce 计算框架只能支持一些特定的计算模式（Map/Reduce），并没有提供一种通用的数据抽象，遇到具体的数据处理工作，要开发相应的 map 和 reduce 函数，开发效率极低。

RDD（Resilient Distributed Dataset，弹性分布式数据集）提供了一个抽象的数据模型，让我们不必担心底层数据的分布式特性，只需将具体的业务应用逻辑表达为一系列可转换的操作（函数），不同 RDD 之间的转换操作还可以形成宽与窄的依赖关系，进而实现管道化，从而避免中间结果的磁盘存储，大大减少数据复制、磁盘 I/O 和序列化开销，并且 Spark 还提供了丰富的 RDD API 方法（算子，如 map、reduce、filter、groupBy 等）。

6.1 RDD 概念与详解

本小节将介绍 RDD 的基本概念，详解 RDD 的属性及关系。

6.1.1 RDD 简介

RDD 是 Spark 中基本的数据抽象，代表一个不可变、可分区、里面的元素可并行计算的集合。

在后面的学习中，我们将会学到两种创建 RDD 的方法：一种是在驱动程序中使用 parallelize 方法加载一个已存在的集合，另一种是在外部存储系统中引用一个数据集。

RDD 将 Spark 的底层细节隐藏了起来，包括自动容错、位置感知、任务调度执行、失败重试等，从而让开发者可以像操作本地集合一样，以函数式编程的方式操作 RDD，进行各种并行业务模型的计算处理。

6.1.2　RDD 的主要属性

我们通过 RDD.scala 的源码文件内容来认识 RDD 的主要属性，如图 6-1 所示。

```
62   * Internally, each RDD is characterized by five main properties:
63   *
64   *  - A list of partitions
65   *  - A function for computing each split
66   *  - A list of dependencies on other RDDs
67   *  - Optionally, a Partitioner for key-value RDDs (e.g. to say that the RDD is hash-partitioned)
68   *  - Optionally, a list of preferred locations to compute each split on (e.g. block locations for
69   *    an HDFS file)
70   *
71   * All of the scheduling and execution in spark is done based on these methods, allowing each RDD
72   * to implement its own way of computing itself. Indeed, users can implement custom RDDs (e.g. for
73   * reading data from a new storage system) by overriding these functions. Please refer to the
74   * <a href="http://people.csail.mit.edu/matei/papers/2012/nsdi_spark.pdf">Spark paper</a>
75   * for more details on RDD internals.
76   */
77  abstract class RDD[T: ClassTag](
```

图 6-1

1. A list of partitions

一组分片 / 一个分区（Partition）列表，即数据集的基本组成单位。对于 RDD 来说，每个分区都会被一个计算任务处理，分区数决定并行度。用户可以在创建 RDD 时指定 RDD 的分区数，如果没有指定，那么会采用默认值。RDD 逻辑上是分区的，每个分区的数据是抽象存在的，计算的时候会通过 compute 函数得到每个分区的数据。如果 RDD 通过已有的文件系统构建，则 compute 函数读取指定文件系统中的数据；如果 RDD 通过其他 RDD 转换而来，则 compute 函数执行转换逻辑。

2. A function for computing each split

一个函数会被作用在每一个分区上。Spark 中 RDD 的计算是以分区为单位的，compute 函数会被作用在每个分区上，如图 6-2 所示。

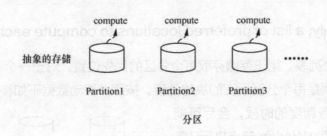

图 6-2

3. A list of dependencies on other RDDs

RDD 的每次转换都会生成一个新的 RDD，所以 RDD 之间就会形成像流水线一样的前后依赖关系。在部分分区数据丢失时，Spark 可以通过这个依赖关系重新计算丢失的分区数据，而不是对 RDD 的所有分区进行重新计算，这是 Spark 容错机制的一部分。

RDD 通过操作算子进行转换，转换得到的新 RDD 包含从其他 RDD 衍生所必需的信息，RDD 之间维护着这种血缘关系，也称为依赖。依赖包括两种：一种是窄依赖，RDD 之间的分区是一一对应的；另一种是宽依赖，下游 RDD 的每个分区与上游 RDD（也称为父 RDD）的每个分区都有关，是多对多的关系，如图 6-3 所示。

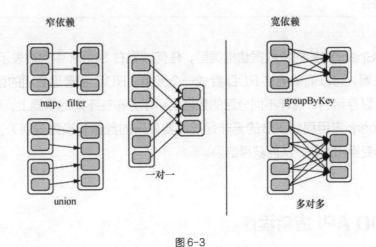

图 6-3

4. Optionally, a Partitioner for key-value RDDs

可选项，对于 key-value 类型的 RDD 会有一个分区器（Partitioner），即 RDD 的分区函数。当前 Spark 中实现了两种类型的分区函数：一种是基于哈希的 HashPartitioner，另外一种是基于范围的 RangePartitioner。只有 key-value 类型的 RDD，才有 Partitioner，非 key-value 类型的 RDD 的 Partitioner 的值是 None。Partitioner 函数不但决定了 RDD 本身的分区数量，也决定了父 RDD Shuffle 输出时的分区数量。

5. Optionally, a list of preferred locations to compute each split on

可选项，一个列表，用于存储存取每个分区的优先位置。对于一个 HDFS 文件来说，这个列表保存的就是每个分区所在的块的位置。按照"移动数据不如移动计算"的理念，Spark 在进行任务调度的时候，会尽可能选择那些存有数据的 Worker 节点执行任务计算。

RDD 是只读的，要想改变 RDD 中的数据，只能在现有 RDD 基础上创建新的 RDD；由一个 RDD 转换到另一个 RDD，可以通过丰富的操作算子（如 map、filter、union、join、reduceByKey）实现，不再像 MapReduce 那样只能写 map 和 reduce，如图 6-4 所示。

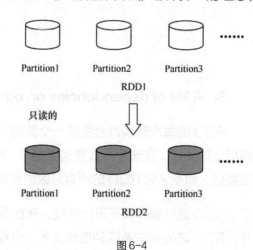

图 6-4

6.1.3 小结

RDD 是 Spark 中的抽象数据结构类型，任何数据在 Spark 中都被表示为 RDD。从编程的角度来看，可以简单地将 RDD 看成一个数组。RDD 和普通数组的区别是，RDD 中的数据是分区存储的，这样不同分区的数据就可以分布在不同的机器上，可以被并行处理。因此，Spark 应用程序所做的无非是把需要处理的数据转换为 RDD，然后对 RDD 进行一系列的变换和操作，从而获得结果。

6.2　RDD API 应用程序

Spark 为我们提供了丰富的 RDD API 操作算子，下面我们从 RDD 的创建开始学习。

用 textFile 方法从文件系统中加载数据来创建 RDD，即由外部存储系统的数据集创建 RDD，包括本地的文件系统，还有所有 Hadoop 支持的数据集，比如 HDFS、Cassandra、HBase 等，以及 Aamzon S3。

```
//HDFS加载数据
val rdd1 = sc.textFile("hdfs://master:9000/user/hadoop/words.txt")
```

```
val rdd2 = sc.textFile("/user/hadoop/words.txt")
```
// 从本地 Linux 文件系统加载数据
```
val lines = sc.textFile("file:///usr/local/spark/mycode/rdd/words.txt")
```

已有的 RDD 经过算子转换成新的 RDD。

```
val rdd4=rdd1.flatMap(_.split(" "))
```

由一个已经存在的 Scala 集合，通过一个数组来创建 RDD。

```
val rdd5 = sc.parallelize(Array(1,2,3,4,5,6,7,8))
```

或者通过一个列表来创建 RDD。

```
val rdd6 = sc.makeRDD(List(1,2,3,4,5,6,7,8))
```

注意，makeRDD 方法底层调用了 parallelize 方法，源码如图 6-5 所示。

```
def makeRDD[T: ClassTag](
    seq: Seq[T],
    numSlices: Int = defaultParallelism): RDD[T] = withScope {
  parallelize(seq, numSlices)
}
```

图 6-5

6.3 RDD 的方法（算子）分类

Spark 为我们提供了两种类型的 RDD 算子。

Transformation（转换）操作：返回一个新的 RDD。在实际的业务处理中，Transformation 的执行将会非常频繁。Spark 也为我们提供了丰富的 Transformation 算子，如 map、filter、flatMap、mapPartitions 等。转换得到的 RDD 是惰性求值的。也就是说，整个转换过程只记录了转换的轨迹，并不会发生真正的计算，只有遇到动作操作时，才会发生真正的计算，从血缘关系源头开始，进行物理上的转换。

Action（动作）操作：返回值不是 RDD，无返回值或返回其他的。常见的 Action 算子有 reduce、collect、count、saveAsTextFile 等。Action 是真正触发计算的地方。Spark 程序执行到 Action 时，才会进行真正的计算：从文件中加载数据，完成一次又一次的 Transformation，最终完成 Action 得到结果。Action 触发了 Spark Job（作业）的执行，应用程序中如果有多个 Action，那么会对应多个 Job，如图 6-6 所示。

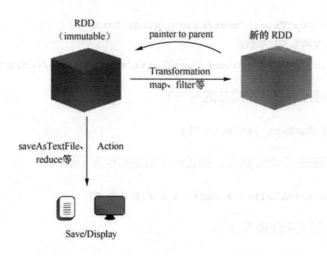

图 6-6

通过前面的学习我们知道 RDD 不实际存储真正要计算的数据,而是记录数据的位置以及数据的转换关系:调用什么方法、传入什么函数等。

RDD 中的所有转换都是惰性求值 / 延迟执行的,也就是说并不会直接计算。只有当发生一个要求返回结果给 Driver 的 Action 时,这些转换才会真正执行。

之所以使用惰性求值 / 延迟执行,是因为这样可以在执行 Action 时对 RDD 操作形成 DAG,以进行阶段的划分和并行优化,这种设计让 Spark 可以更加有效率地运行。相关内容我们将在后面进行详细的介绍。下面我们来看 Spark 提供了哪些具体的 Transformation 和 Action 算子,我们先来了解和认识一下这些算子。

6.3.1 Transformation 算子

表 6-1 列出了 Spark 框架提供的 Transformation 算子及其含义。

表 6-1

Transformation 算子	含义
map(func)	返回一个新的 RDD,该 RDD 由每一个输入元素经过 func 函数转换后组成
filter(func)	返回一个新的 RDD,该 RDD 由经过 func 函数计算后返回值为 true 的输入元素组成
flatMap(func)	类似于 map,但是每一个输入元素可以被映射为 0 或多个输出元素(所以 func 应该返回一个序列,而不是单一元素)

续表

Transformation 算子	含义
mapPartitions(func)	类似于 map，但独立地在 RDD 的每一个分区上运行，因此在类型为 T 的 RDD 上运行时，func 的函数类型必须是 Iterator[T] => Iterator[U]
mapPartitionsWithIndex(func)	类似于 mapPartitions，但 func 带有一个整数参数，表示分区的索引值，因此在类型为 T 的 RDD 上运行时，func 的函数类型必须是 (Int, Iterator[T]) => Iterator[U]
sample(withReplacement, fraction, seed)	根据 fraction 指定的比例对数据进行采样，可以选择是否使用随机数进行替换，seed 用于指定随机数生成器种子
union(otherDataset)	对源 RDD 和参数 RDD 求并集后返回一个新的 RDD
intersection(otherDataset)	对源 RDD 和参数 RDD 求交集后返回一个新的 RDD
distinct([numTasks]))	对源 RDD 进行去重后返回一个新的 RDD
groupByKey([numTasks])	在一个 (K,V) 类型的 RDD 上调用，返回一个 (K, Iterator[V]) 类型的 RDD
reduceByKey(func, [numTasks])	在一个 (K,V) 类型的 RDD 上调用，返回一个 (K,V) 类型的 RDD，使用指定的 reduce 函数，将相同 key 的值聚合到一起。与 groupByKey 类似，reduce 任务的个数可以通过第二个可选的参数来设置
sortByKey([ascending], [numTasks])	在一个 (K,V) 类型的 RDD 上调用，K 必须实现 Ordered 接口，返回一个按照 key 进行排序的 (K,V) 类型的 RDD
sortBy(func,[ascending], [numTasks])	与 sortByKey 类似，但是更灵活
join(otherDataset, [numTasks])	在类型为 (K,V) 和 (K,W) 的 RDD 上调用，返回一个相同 key 对应的所有元素对在一起的 (K,(V,W)) 的 RDD
cogroup(otherDataset, [numTasks])	在类型为 (K,V) 和 (K,W) 的 RDD 上调用，返回一个 (K,(Iterable<V>,Iterable<W>)) 类型的 RDD
cartesian(otherDataset)	笛卡儿积
pipe(command, [envVars])	对 Rdd 进行管道操作
coalesce(numPartitions)	减少 RDD 的分区数到指定值。在过滤大量数据之后，可以执行此操作
repartition(numPartitions)	重新给 RDD 分区

6.3.2 Action 算子

表 6-2 列出了 Spark 框架为我们提供的 Action 算子及其含义。

表 6-2

Action 算子	含义
reduce(func)	通过 func 函数聚集 RDD 中的所有元素，这个操作必须是可交换且可并联的
collect	在驱动程序中，以数组的形式返回数据集的所有元素
count	返回 RDD 的元素个数
first	返回 RDD 的第一个元素（类似于 take(1)）
take(n)	返回一个由数据集的前 n 个元素组成的数组
takeSample(withReplacement,num, [seed])	返回一个数组，该数组由从数据集中随机采样的 num 个元素组成，可以选择是否用随机数替换不足的部分，seed 用于指定随机数生成器种子
takeOrdered(n, [ordering])	返回自然顺序或者自定义顺序的前 n 个元素
saveAsTextFile(path)	将数据集的元素以文本文件的形式保存到 HDFS 或者其他支持的文件系统。对于每个元素，Spark 将会调用 toString 方法将它转换为文件中的文本
saveAsSequenceFile(path)	将数据集的元素以 Hadoop sequencefile 的形式保存到指定的目录下，可以是 HDFS 或者其他 Hadoop 支持的文件系统
saveAsObjectFile(path)	将数据集的元素以 Java 序列化的方式保存到指定的目录下
countByKey	针对 (K,V) 类型的 RDD，返回一个 (K,Int) 类型的 map，表示每一个 key 对应的元素个数
foreach(func)	在数据集的每一个元素上，运行函数 func 进行更新
foreachPartition(func)	在数据集的每一个分区上，运行函数 func

6.4 基础练习

在对 Transformation 和 Action 有了基本的认识之后，下面我们就来做一些基础练习。

首先启动 Spark 集群环境（命令是 /export/servers/spark/sbin/start-all.sh），当 master 节点出现 Master 进程，slave 和 slave1 节点分别出现 Worker 进程的时候，证明 Spark 集群环境启动成功。

接着启动 spark-shell，即 Spark 框架自带的交互式 Shell 程序，命令如下。

```
/export/servers/spark/bin/spark-shell \
--master spark://master:7077 \
--executor-memory 1g \
--total-executor-cores 2
```

或者（以本地模式启动）

```
/export/servers/spark/bin/spark-shell
```

通过执行以上命令，将打开 Spark 交互式编程的窗口，如图 6-7 所示，在方框的位置就可以写 Spark 应用程序进行交互式编程操作。

```
Spark context Web UI available at http://192.168.52.102:4040
Spark context available as 'sc' (master = spark://master:7077, app id = app-20210829211450-0000).
Spark session available as 'spark'.
Welcome to
      ____              __
     / __/__  ___ _____/ /__
    _\ \/ _ \/ _ `/ __/  '_/
   /___/ .__/\_,_/_/ /_/\_\   version 2.2.0
      /_/

Using Scala version 2.11.8 (Java HotSpot(TM) 64-Bit Server VM, Java 1.8.0_144)
Type in expressions to have them evaluated.
Type :help for more information.

scala>
```

图 6-7

6.4.1 实现 WordCount 案例

首先在 HDFS 中查看 words.txt 中的数据，如图 6-8 所示。

```
[hadoop@master ~]$ hadoop fs -cat /wordcount/input/words.txt
hello me hadoop
hello you hadoop
hello her hadoop
hello me spark
hello you spark
hello her hello
hello me spark
hello you hadoop
hello her hadoop
hello me spark
hello you hbase
hello her hbase
```

图 6-8

假如我们需要统计在这个文件中"hello"这个单词出现了多少次,以及"hadoop"这个单词出现了多少次,等等。通过 Spark 提供的 Transformation 和 Action 两种类型的算子就可以实现这个功能。

首先,通过 textFile 将数据从 HDFS 中加载进来,然后用 flatMap 按空格进行切分,映射成一个个单词,再通过 map 映射成 key-value,最后通过 reduceByKey 根据 key 相同的汇聚统计出各个单词频次。代码如下:

```
val res = sc.textFile("hdfs://master:9000/wordcount/input/words.txt")
.flatMap(_.split(" ")).map((_, 1)).reduceByKey(_+_)
```

代码运行结果如图 6-9 所示。

```
scala> res.collect
res1: Array[(String, Int)] = Array((hello,13), (me,4), (spark,4), (you,4), (hadoop,5), (her,4), (hbase,2))
```

图 6-9

6.4.2 创建 RDD

通过调用 SparkContext 中的 parallelize 方法创建 rdd1,代码如下。

```
val rdd1 = sc.parallelize(List(5,6,4,7,3,8,2,9,1,10))
```

运行结果如下所示。

```
scala> rdd1.collect
res0: Array[Int] = Array(5, 6, 4, 7, 3, 8, 2, 9, 1, 10)
```

通过调用 SparkContext 中的 makeRDD 方法创建 rdd2,代码如下。

```
val rdd2 = sc.makeRDD(List(1,2,3,4,5,6,7,8,9,10))
```

运行结果如下所示。

```
scala> rdd2.collect
res1: Array[Int] = Array(1, 2, 3, 4, 5, 6, 7, 8, 9, 10)
```

我们还可以查看 RDD 的分区数,代码如下。

```
//没有指定分区数,默认值是2
scala> sc.parallelize(List(1,2,3,4,5,6,7,8,9,10)).partitions.length
res5: Int = 2
```

```
//指定分区数为3
sc.parallelize(List(5,6,4,7,3,8,2,9,1,10),3).partitions.length
//没有指定分区数，默认值是2
scala>sc.textFile("hdfs://master:9000/wordcount/input/words.txt").partitions.length
res7: Int = 2
```

RDD 分区的原则是使得分区数尽量等于集群中的 CPU 核数，这样可以充分利用 CPU 的计算资源。但是在实际中为了更加充分地使用 CPU 的计算资源，会把并行度设置为 CPU 核数的 2～3 倍，RDD 分区数和启动时指定的 CPU 核数、调用方法时指定的分区数（如文件本身的分区数）有关系。

6.4.3 map

map 是对 RDD 中的每一个元素进行操作并返回操作的结果，通过并行化生成 RDD，代码如下。

```
val rdd1 = sc.parallelize(List(1, 2, 3, 4, 5, 6, 7, 8, 9, 10))
```

代码运行结果如下所示。

```
scala> rdd1.collect
res9: Array[Int] = Array(1, 2, 3, 4, 5, 6, 7, 8, 9, 10)
//对rdd1里的每一个元素乘2，然后返回新的数组
val rdd2 = rdd1.map(_ * 2)
//collect方法用于收集结果，属于Action操作
rdd2.collect
```

我们已经看到 rdd2 中的每一个元素是 rdd1 中的每一个元素的两倍。

```
scala> rdd2.collect
res8: Array[Int] = Array(2, 4, 6, 8, 10, 12, 14, 16, 18, 20)
```

6.4.4 filter

filter 函数中返回 true 的元素被留下，返回 false 的元素被过滤。如下所示，rdd1 中含有 10 个元素，如果想把大于等于 7 的元素从 rdd1 中过滤出来，操作代码如下。

```
val rdd1 = sc.parallelize(List(1, 2, 3, 4, 5, 6, 7, 8, 9, 10))
//调用filter将rdd1中大于等于7的元素过滤出来
```

```
val rdd2 = rdd1.filter(_ >= 7)
//collect方法用于收集结果,属于Action操作
scala> rdd2.collect
res10: Array[Int] = Array(7, 8, 9, 10)
```

6.4.5　flatMap

flatMap 是对 RDD 中的每一个元素先进行 map 再压平,最后返回操作的结果。rdd1 中有 3 个元素,分别是"ａｂｃ""ｄｅｆ""ｈｉｊ"。现在,我们需要把这 3 个元素中的每个字母都提出来形成一个集合"ａｂｃｄｅｆｈｉｊ"。可以通过如下操作得出结果。

```
val rdd1 = sc.parallelize(Array("a b c", "d e f", "h i j"))
//将rdd1里面的每一个元素先按空格切分再压平
val rdd2 = rdd1.flatMap(_.split(" "))
rdd2.collect
//Array[String] = Array(a, b, c, d, e, f, h, i, j)
```

6.4.6　sortBy

创建一个 rdd1,代码如下。

```
val rdd1 = sc.parallelize(List(5, 6, 4, 7, 3, 8, 2, 9, 1, 10))
```

对 rdd1 中的元素进行排序,其中 x=>x 表示按照元素本身进行排序,true 表示升序,false 表示降序。

```
//升序
scala> val rdd2 = rdd1.sortBy(x=>x, true).collect
rdd2: Array[Int] = Array(1, 2, 3, 4, 5, 6, 7, 8, 9, 10)
//降序
scala> val rdd2 = rdd1.sortBy(x=>x, false).collect
rdd2: Array[Int] = Array(10, 9, 8, 7, 6, 5, 4, 3, 2, 1)
    val rdd2 = rdd1.sortBy(x=>x+"", true)    //x=>x+"" 表示按照x的字符串形式排序结果为
字典顺序
            scala> val rdd2 = rdd1.sortBy(x=>x+"", false).collect
rdd2: Array[Int] = Array(9, 8, 7, 6, 5, 4, 3, 2, 10, 1)
```

6.4.7　交集、并集、差集、笛卡尔积

下面我们创建两个类型一致的 RDD，代码如下。

```
val rdd1 = sc.parallelize(List(5, 6, 4, 3))
val rdd2 = sc.parallelize(List(1, 2, 3, 4))
//union不会去重
scala>val rdd3 = rdd1.union(rdd2).collect
res11: Array[Int] = Array(5, 6, 4, 3, 1, 2, 3, 4)
//去重，将rdd3中的重复元素去掉
ala> rdd3.distinct
res13: Array[Int] = Array(5, 6, 4, 3, 1, 2)
//求交集，将rdd1和rdd2中都有的元素提取出来
scala> val rdd4 = rdd1.intersection(rdd2).collect
rdd4: Array[Int] = Array(4, 3)
//求差集，将rdd1中的元素与rdd2中的元素重复的去掉
scala> val rdd5 = rdd1.subtract(rdd2).collect
rdd5: Array[Int] = Array(6, 5)
//求笛卡尔积，创建学生rdd1和课程rdd2
val rdd1 = sc.parallelize(List("Jack", "Tom"))//学生
val rdd2 = sc.parallelize(List("Java", "Python", "Scala"))//课程
val rdd3 = rdd1.cartesian(rdd2)//表示所有学生的所有选课情况
rdd3.collect
Array[(String, String)] = Array((Jack,Java), (Jack,Python), (Jack,Scala),
(Tom,Java), (Tom,Python), (Tom,Scala))
```

6.4.8　groupByKey

groupByKey 算子的功能是，对具有相同 Key 的值进行分组。比如，对键值对 ("spark",1)、("spark",2)、("hadoop",3) 和 ("hadoop",5) 采用 groupByKey 后得到的结果是 ("spark",(1,2)) 和 ("hadoop",(3,5))。

```
//按key进行分组
val rdd1= sc.parallelize(Array(("tom",1), ("jerry",2), ("kitty",3), ("jerry",9),
("tom",8), ("shuke",7), ("tom",2)))
//对rdd1中有相同Key的值进行分组
val rdd2=rdd1.groupByKey
rdd2.collect
```

运行结果如下所示。

```
Array[(String, Iterable[Int])] = Array((tom,CompactBuffer(1, 8, 2)), (jerry,-
CompactBuffer(2, 9)), (shuke,CompactBuffer(7)), (kitty,CompactBuffer(3)))
```

6.4.9　groupBy

groupBy 算子的功能是根据指定的函数中的规则对 key 进行分组。创建一个 intRdd，要想把其中的偶数和奇数分开，代码如下。

```
val intRdd = sc.parallelize(List(1,2,3,4,5,6))
val result = intRdd.groupBy(x=>{if(x%2 == 0) "even" else "odd"}).collect
//intRdd中的奇数和偶数已经被分开了
Array[(String, Iterable[Int])] = Array((even,CompactBuffer(2, 4, 6)), (odd,-
CompactBuffer(1, 3, 5)))
```

6.4.10　reduce

reduce 算子的功能是对集合中的元素进行聚合运算。例如，我们想对 rdd1 中的 5 个元素进行累加运算，代码如下。

```
val rdd1 = sc.parallelize(List(1, 2, 3, 4, 5))
//reduce聚合
scala> val result = rdd1.reduce(_+_)
//第一个 "_" 代表上次运算的结果，第二个 "_" 代表这次进来的元素
result: Int = 15
```

6.4.11　reduceByKey

reduceByKey 算子的功能是执行 Transformation；reduceByKey(func) 算子的功能是使用 func 函数合并具有相同 key 的值。比如，执行 reduceByKey((a,b) => a+b)，有 4 个键值对 ("spark",1)、("spark",2)、("hadoop",3) 和 ("hadoop",5)，对具有相同 key 的键值对进行合并后的结果就是 ("spark",3)、("hadoop",8)。可以看出，(a,b) => a+b 这个 Lambda 表达式中，a 和 b 都是指 value，比如，对于两个具有相同 key 的键值对 ("spark",1)、("spark",2)，a 就是 1，b 就是 2。

```
val rdd1 = sc.parallelize(List(("tom", 1), ("jerry", 3), ("kitty", 2),
("shuke", 1)))
```

```
val rdd2 = sc.parallelize(List(("jerry", 2), ("tom", 3), ("shuke", 2), ("kitty", 5)))
val rdd3 = rdd1.union(rdd2)  //并集
rdd3.collect
/*Array[(String, Int)] = Array((tom,1), (jerry,3), (kitty,2), (shuke,1), (jerry,2),
(tom,3), (shuke,2), (kitty,5))*/
//按key进行聚合
val rdd4 = rdd3.reduceByKey(_ + _)
rdd4.collect
//Array[(String, Int)] = Array((tom,4), (jerry,5), (shuke,3), (kitty,7))
```

6.4.12 repartition

使用 repartition 算子在把处理结果保存到 HDFS 之前可以减少分区数，这样会起到合并小文件的效果，可以使数据保存到 HDFS 上的效率有所提升。repartition 算子的功能是改变 RDD 的分区数。

```
val rdd1 = sc.parallelize(1 to 10,3)  //指定3个分区
//利用repartition改变rdd1的分区数
//减少分区
val rdd2 = rdd1.repartition(2).partitions.length  //新生成的rdd2的分区数为2
rdd1.partitions.length  //原来的rdd1的分区数不变
```

在将处理结果保存到 HDFS 之前进行重分区，分区数为 1，那么保存在 HDFS 上的结果文件只有一个，代码如下。

```
sc.textFile("hdfs://master:9000/wordcount/input/words.txt")
.flatMap(_.split(" ")).map((_,1)).reduceByKey(_+_).repartition(1)
.saveAsTextFile("hdfs://node01:8020/wordcount/output5")
```

接下来介绍一些一目了然的算子，这里仅给出代码，不赘述。

6.4.13 count

count 算子的功能是统计集合中元素的个数。

```
val rdd1 = sc.parallelize(List(6,1,2,3,4,5), 2)
rdd1.count  // 统计结果为6个元素
```

6.4.14 top

top 算子的功能是取出最大的前几个元素。

```
val rdd1 = sc.parallelize(List(3,6,1,2,4,5))
rdd1.top(2)   //结果为6、5
```

6.4.15 take

take 算子的功能是按照原来的顺序取前 n 个元素。

```
val rdd1 = sc.parallelize(List(6,1,2,3,4,5), 2)
rdd1.take(2)   //结果为6、1
rdd1.sortBy(x=>x,true).take(2)     //结果为1、2
```

6.4.16 first

first 算子的功能是按照原来的顺序取第一个元素。

```
val rdd1 = sc.parallelize(List(6,1,2,3,4,5), 2)
rdd1.first   //结果为6
```

6.4.17 keys、values

创建 rdd1，代码如下。

```
val rdd1 = sc.parallelize(List("dog", "tiger", "lion", "cat", "panther", "eagle"), 2)
//按集合中元素字符串的长度进行map映射
val rdd2 = rdd1.map(x => (x.length, x))
rdd2.collect   //获取结果
Array[(Int, String)] = Array((3,dog), (5,tiger), (4,lion), (3,cat), (7,panther), (5,eagle))
```

使用算子 keys 来获取 rdd2 中所有 key-value 元素的 key，代码如下。

```
rdd2.keys.collect
//Array[Int] = Array(3, 5, 4, 3, 7, 5)
```

使用算子 values 来获取 rdd2 中所有 key-value 元素的 value，代码如下。

```
rdd2.values.collect
//Array[String] = Array(dog, tiger, lion, cat, panther, eagle)
```

6.4.18 案例

创建 rdd，代码如下所示。

```
val rdd = sc.parallelize(Array(("spark",2),("hadoop",6),("hadoop",4),("spark",6)))
```

其中，key 表示图书名称，value 表示某天图书销量，请计算每个 key 对应的平均值，也就是计算每种图书每天的平均销量。最终结果是 ("spark",4)、("hadoop",5)。下面我们通过 Spark 提供的 groupByKey 和 map 来实现这个需求，代码如下。

```
//加载图书销量数据
    val rdd = sc.parallelize(Array(("spark",2),("hadoop",6),("hadoop",4),
("spark",6)))
    //通过算子groupByKey，对具有相同key的值进行分组
    val rdd2 = rdd.groupByKey()
    rdd2.collect
    //Array[(String, Iterable[Int])] = Array((spark,CompactBuffer(2, 6)),
    (hadoop,CompactBuffer(6, 4)))
    //计算每种图书每天的平均销量
    val rdd3 = rdd2.map(t=>(t._1,t._2.sum /t._2.size))
    //t._1表示图书名称，t._2.sum用于累加图书的销量，t._2.size相当于统计销量批次
    rdd3.collect
    //Array[(String, Int)] = Array((spark,4), (hadoop,5))
```

6.5 实战案例

6.5.1 统计平均年龄

假设需要统计 1000 万城市人口的平均年龄，这些年龄信息都存储在 sample1.dat 文件里，数据格式如图 6-10 所示，第一列是 ID，第二列是年龄，数据共 1000 万条。

```
1  16
2  74
3  51
4  35
5  44
6  95
7  5
8  29
10 60
11 13
12 99
13 7
14 26
```

图 6-10

1. 第一种方法

将文件加载到 RDD，进行数据有效性过滤，并将其切分为 (年龄 ,1.0) 的数据映射格式，代码如下。

```
//1.加载数据
scala> val fileRDD = sc.textFile("/datas/sample1.dat").filter(_.split(" ").size == 2).map(_.split(" ")).map(x=>(x(1).toDouble, 1.0))
fileRDD: org.apache.spark.rdd.RDD[(Double, Double)] = MapPartitionsRDD[17] at map at <console>:30

//2.将数据缓存
scala> fileRDD.persist(StorageLevel.MEMORY_AND_DISK)
res6: fileRDD.type = MapPartitionsRDD[17] at map at <console>:30

//3.显示数据切分的结果
scala> fileRDD.take(10)
res7: Array[(Double, Double)] = Array((15.0,1.0), (32.0,1.0), (55.0,1.0), (83.0,1.0), (85.0,1.0), (80.0,1.0), (90.0,1.0), (95.0,1.0), (2.0,1.0), (74.0,1.0))
//累加年龄
scala> def arrAdd(x: Array[Double], y:Array[Double]): Array[Double] = x.zip(y).map(x=>x._1+x._2)
arrAdd: (x: Array[Double], y: Array[Double])Array[Double]
scala> val (sums, count) = fileRDD.map(x=>(Array(x._1, x._2), 1)).reduce((x,y) => (arrAdd(x._1, y._1), x._2+y._2))
sums: Array[Double] = Array(4.95010657E8, 1.0E7)
count: Int = 10000000
scala> val result = sums(0) / count
result: Double = 49.5010657
scala> val result = sums(0) / sums(1)
result: Double = 49.5010657
```

2. 第二种方法

加载数据文件，做数据的清洗，之后参与业务处理，代码如下。

```
scala> val fileRDD = sc.textFile("/datas/sample1.dat").filter(_.split(" ").
size == 2).map(_.split(" ")).map(x=>(x(1).toDouble, 1.0))
fileRDD: org.apache.spark.rdd.RDD[(Double, Double)] = MapPartitionsRDD[33] at map
at <console>:30

scala> val (sums, count) = fileRDD.reduce((x, y)=>(x._1 + y._1, x._2 + y._2))
sums: Double = 4.95010657E8
count: Double = 1.0E7

scala> val result = sums / count
result: Double = 49.5010657
```

3. 第三种方法

通过 main 方法完成数据加载、清洗及业务数据分析。

```
object AvgAge {
  def main(args: Array[String]): Unit = {
    //获取配置信息
    val conf = new SparkConf().setAppName("AvgAge").setMaster("local")
    //创建SparkContext对象
    val sc = new SparkContext(conf)
    //记载数据
    val file = sc.textFile("D:/java/Spark/day11_spark/sample1.dat")
    //数据处理
    val rdd = file.map(_.split(" ")(1).toDouble).map((_,1.0))
    val (sums,count)=rdd.reduce((x,y)=>(x._1+x._2,y._1+y._2))
    //获取结果
    val avg = (sums/count).round
    print("平均年龄为: "+avg)
  }
}
```

运行结果如下。

平均年龄为: 49.5010657

6.5.2 统计人口信息

对某个省人口的性别、身高进行统计,需要统计出男女人数,男性中的最高和最低身高,以及女性中的最高和最低身高。源文件有 3 列,分别为 ID、性别和身高(单位为 cm），数据格式如图 6-11 所示。

```
1 M 174
2 F 165
3 M 172
4 M 180
5 F 160
6 F 162
7 M 172
```

图 6-11

1. 第一种方法

通过交互式编程的方法实现，代码如下。

```
scala> val allFileRDD = sc.textFile("/datas/sample_people_info.dat").filter(_.split(" ").size == 3).map(_.split(" ")).map(x=>(x(1), x(2)))
allFileRDD: org.apache.spark.rdd.RDD[(String, String)] = MapPartitionsRDD[43] at map at <console>:30

scala> allFileRDD.persist(StorageLevel.MEMORY_AND_DISK)
res19: allFileRDD.type = MapPartitionsRDD[43] at map at <console>:30

scala> val maleRDD = allFileRDD.filter(x=>(x._1 == "M"))
maleRDD: org.apache.spark.rdd.RDD[(String, String)] = MapPartitionsRDD[44] at filter at <console>:31

scala> maleRDD.persist(StorageLevel.MEMORY_AND_DISK)
res21: maleRDD.type = MapPartitionsRDD[44] at filter at <console>:31

scala> val femaleRDD = allFileRDD.filter(x=>(x._1 == "F"))
femaleRDD: org.apache.spark.rdd.RDD[(String, String)] = MapPartitionsRDD[45] at filter at <console>:31

scala> femaleRDD.persist(StorageLevel.MEMORY_AND_DISK)
res22: femaleRDD.type = MapPartitionsRDD[45] at filter at <console>:31

// 男性人数总和
scala> maleRDD.count()
// 女性人数总和
scala> femaleRDD.count()
// 男性最高身高
scala> maleRDD.map(x=>(x._2.toDouble)).sortBy(x => x,false).first()
(F,229)
```

```
// 男性最低身高
scala> maleRDD.map(x=>(x._2.toDouble)).sortBy(x => x,true).first()
(F,100)
// 女性最高身高
scala> femaleRDD.map(x=>(x._2.toDouble)).sortBy(x => x,false).first()
(F,229)
// 女性最低身高
scala> femaleRDD.map(x=>(x._2.toDouble)).sortBy(x => x,true).first()
(M,100)
```

2. 第二种方法

通过 main 方法实现，代码如下。

```
object PeopleCount {
  def main(args: Array[String]): Unit = {
    val conf = new SparkConf().setAppName("Population").setMaster("local")
    val sc = new SparkContext(conf)
    val file = sc.textFile("file:///D:/java/Spark/day11_spark /sample_people_info.dat").cache()
    val rdd = file.map(_.split(" ")).map(x=>(x(1),x(2))).cache()
    val rdd1 = rdd.groupByKey().map(x=>(x._1,x._2.max))
    val rdd3 = rdd.groupByKey().map(x=>(x._1,x._2.min))

    val rdd2 = file.map(_.split(" ")(1)).map((_,1))
    val sum = rdd2.reduceByKey(_+_)
    print("男女最高身高: "+rdd1.collect().mkString)
    print("男女最低身高: "+rdd3.collect().mkString)
    print("男女人数: "+sum.collect().mkString)
  }
}
```

将 jar 包放到 Spark 集群中运行。

/root/IdeaProjects/nyFirstMaven/out/artifacts/nyFirstMaven_jar2/nyFirstMaven.jar

完成之后，提交作业。

```
[root@master txt]# spark-submit --class PeopleCount --executor-memory 512M /root/IdeaProjects/nyFirstMaven/out/artifacts/nyFirstMaven_jar2/nyFirMaven.jar hdfs://master:9000/PeopleInfo.txt
```

运行结果如图 6-12 所示。

```
Male counts is 49997895
Female counts is 50002105
Lowset Male:100
Lowset Female:100
Highest Male:229
Highest Female:229
2019-08-18 08:29:12 INFO  SparkContext:54 - Invoking stop() from shutdown hook
2019-08-18 08:29:12 INFO  AbstractConnector:318 - Stopped Spark@d1f74b8{HTTP/1.1,[http/1.1]}{0.0.0.0:404
2019-08-18 08:29:12 INFO  SparkUI:54 - Stopped Spark web UI at http://master:4040
2019-08-18 08:29:12 INFO  StandaloneSchedulerBackend:54 - Shutting down all executors
2019-08-18 08:29:12 INFO  CoarseGrainedSchedulerBackend$DriverEndpoint:54 - Asking each executor to shut
```

图 6-12

6.5.3　在 IDEA 中实现 WordCount 案例

按照 count 值降序显示前 50 行数据，将单词转换成小写，去除标点符号，去除停用词。

在处理自然语言数据（或文本）之前或之后会自动过滤掉某些字或词，这些字或词被称为停用词（Stop Word）。这些停用词都是人工输入、非自动化生成的，生成后的停用词会形成一个停用词表。但是，并没有一个明确的停用词表能适用于所有工具。

通常意义上的停用词大致分为两类：一类是没有什么实际含义的词，比如 the 等，但是对于搜索引擎来说，当用户要搜索 The Who、The The 或 Take The 等复合名词时，停用词的使用就会导致问题；另一类词，比如 want 等，搜索引擎无法保证能够给出真正相关的搜索结果，难以帮助用户缩小搜索范围，同时还会降低搜索的效率。这些词通常会被移除，从而提高搜索性能。

- 分析：移除停用词（停用词可大体划分类别，可以使用 Set 进行存储）、去除标点符号之后将单词字母转换为小写，统计转换后的小写单词个数。
- 数据文件：Ugly Duckling.txt。
- 数据格式如图 6-13 所示。

```
Ugly Duckling.txt - 记事本
文件(F)  编辑(E)  格式(O)  查看(V)  帮助(H)
Ugly Duckling
The countryside was lovely. It was summer. The wheat was golden and the oats were still green. The hay was stacked in the low-lying meadows.
In the depths of a forest a duck was sitting in her nest. Her little ducklings were about to be hatched.
At last one egg after another began to crack." Cheep, cheep!" the ducklings said." Quack, quack!" said the duck. " How big the world is!" said all
But the biggest egg was still there. And then she settled herself on the nest again.
"Well, how are you getting on?" said an old duck who came to pay her a visit." This egg is taking such a long time," answered the sitting duck."
The shell will not crack, but the others are the finest ducklings. They are like their father."
"let me look at the egg which won't crack," said the old duck." You may be sitting on a turkey's egg! I have been cheated like that once. Yes. it's
```

图 6-13

演示代码如下。

```scala
object WordCount {
  def main(args: Array[String]): Unit = {
    val conf = new SparkConf().setAppName("WordCount").setMaster("local")
    val sc = new SparkContext(conf)
    val file = sc.textFile("file:///D:/Ugly Duckling.txt")
    val rdd1 = file.flatMap(_.split(" ")).map(x=>x.toLowerCase).filter(x=>x.equals("the")==false)
            .filter(x=>x.equals("is")==false).filter(x=>x.equals("at")==false).filter(x=>x.equals ("which")==false)
            .filter(x=>x.equals("on")==false).filter(x=>x.equals("'t")==false).filter(x=>x.equals("!")==false)
     .filter(x=>x.equals("?")==false).filter(x=>x.equals(",")==false).filter(x=>x.equals("won't")==false)
     .filter(x=>x.equals("?\"")==false).map(x=>
    if(x.startsWith("\""))
      x.substring(0,x.indexOf("\""))
       else if(x.endsWith("."))
         x.substring(x.indexOf("."),x.length)
       else if(x.endsWith(","))
         x.substring(x.indexOf(","),x.length)
       else if(x.endsWith("\""))
         x.substring(x.indexOf("\""),x.length)
       else if(x.endsWith("?"))
         x.substring(x.indexOf("?"),x.length)
       else if(x.endsWith("!"))
         x.substring(x.indexOf("!"),x.length)
       else if(x.endsWith("'s"))
         x.substring(x.indexOf("'s"),x.length)
       else x
    )
    val rdd2 = rdd1.map(x=>{(x,1)}).reduceByKey(_+_)
    print(rdd2.collect().mkString)
  }
}
```

运行结果如下。

(branches,1)(next,1)(it,5)(ducks,4)(than,2)(others,2)(hay,1)(greater,1)(full,1)……

6.5.4 小结

RDD 的算子分为两类,一类是 Transformation,另一类是 Action。那么如何区分 Transformation 和 Action 呢?返回值是 RDD 的为 Transformation,其特点是延迟执行/懒执行/惰性执行。返回值不是 RDD(如返回值的数据类型为 Unit、Array、Int 等)的为 Action。需要注意的是,RDD 不实际存储真正要计算的数据,只记录 RDD 的转换关系(调用什么方法。传入什么函数。依赖哪些 RDD。分区器是什么。数量块来源机器列表)。RDD 中的所有 Transformation 操作都是延迟执行(懒执行)的,也就是说并不会直接计算。只有当发生 Action 操作的时候,这些 Transformation 操作才会真正执行。

6.6 RDD 持久化缓存

在实际开发中某些 RDD 的计算或转换可能比较耗费时间,如果这些 RDD 后续还会被频繁使用,那么可以将这些 RDD 进行持久化缓存,这样下次使用的时候就不用再重新计算,从而提高应用程序运行的效率。

可以说,持久化缓存是 Spark 构建迭代式算法和快速交互式查询的关键因素。

代码演示

使用 cache 方法将一个 RDD 标记为持久化缓存,之所以说"标记为持久化缓存",是因为出现 cache 语句的地方,并不会马上计算生成 RDD 并把它持久化缓存,而是要等到第一个 Action 操作触发真正的计算以后,才会对计算结果进行持久化缓存。

下面我们通过代码来演示持久化缓存的应用。首先启动 Spark 集群,代码如下。

```
/export/servers/spark/sbin/start-all.sh
```

然后启动 spark-shell,代码如下。

```
/export/servers/spark/bin/spark-shell \
--master spark://master:7077 \
--executor-memory 1g \
--total-executor-cores 2
```

若出现如图 6-14 所示的结果,证明集群环境启动成功。

```
Spark context Web UI available at http://192.168.52.102:4040
Spark context available as 'sc' (master = spark://master:7077, app id = app-2021
0831062009-0000).
Spark session available as 'spark'.
Welcome to
      ____              __
     / __/__  ___ _____/ /__
    _\ \/ _ \/ _ `/ __/  '_/
   /___/ .__/\_,_/_/ /_/\_\   version 2.2.0
      /_/

Using Scala version 2.11.8 (Java HotSpot(TM) 64-Bit Server VM, Java 1.8.0_144)
Type in expressions to have them evaluated.
Type :help for more information.

scala>
```

图 6-14

将一个 RDD 持久化缓存，这样就可以在后续操作中直接从缓存中获取 RDD，代码如下。

```
//加载HDFS中的words.txt文件数据
val rdd1 = sc.textFile("hdfs://master:9000/wordcount/input/words.txt")
//通过RDD的Transformation算子得出词频统计的计算预统计
val rdd2 = rdd1.flatMap(x=>x.split(" ")).map((_,1)).reduceByKey(_+_)
rdd2.cache   //持久化缓存，此时rdd2并没有真正进行持久化缓存，只是被标记出来
rdd2.sortBy(_._2,false).collect
//触发Action，会去读取HDFS中的文件，rdd2会真正进行持久化缓存
rdd2.sortBy(_._2,false).collect
//触发Action，会去读取缓存中的数据，执行速度比之前快，因为rdd2已经持久化缓存到内存中了
```

6.7 持久化缓存 API 详解

缓存是指将计算结果写入不同的介质，用户可自定义存储级别（存储级别定义了缓存存储的介质，目前支持内存、堆外内存和磁盘）。

6.7.1 persist 方法和 cache 方法

Spark 提供了两个持久化缓存的方法：persist 方法和 cache 方法。RDD 通过 persist 或 cache 方法将前面的计算结果缓存，但并不是在这两个方法被调用时立即缓存，只是做了一个持久化缓存标记，当触发后面的操作时，该 RDD 将被缓存在计算节点的内存中，并供后面重复使用。

因此，通过 persist 或 cache 方法可以标记一个要被持久化缓存的 RDD，一旦首次被触发，该 RDD 将会被缓存在计算节点的内存中并重用。

考虑到什么时候该缓存数据，需要对空间和速度进行权衡，并且垃圾回收开销的问题

也会让情况变得更复杂。一般情况下，如果多个 Action 需要用到某个 RDD，而它的计算代价又很高，那么应该把这个 RDD 缓存起来。

假设首先进行了 RDD0 → RDD1 → RDD2 的计算作业，在计算结束时，RDD1 就已经缓存在系统中了，在进行 RDD0 → RDD1 → RDD3 的计算作业时，由于 RDD1 已经缓存在系统中，因此 RDD0 → RDD1 的转换不会重复执行，计算作业只需进行 RDD1 → RDD3 的计算，因此计算速度可以得到很大提升，如图 6-15 所示。

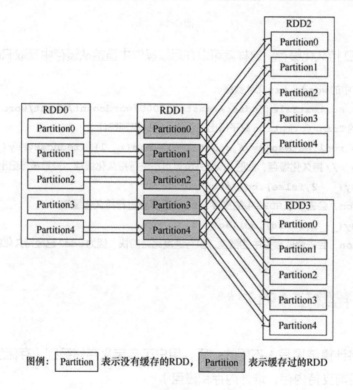

图 6-15

通过查看 RDD 的源码发现 cache 方法最终调用了 persist 方法（默认只存在内存中），如图 6-16 所示。

```
204    /**
205     * Persist this RDD with the default storage level (`MEMORY_ONLY`)//只存在内存中
206     */
207    def persist(): this.type = persist(StorageLevel.MEMORY_ONLY)
208
209    /**
210     * Persist this RDD with the default storage level (`MEMORY_ONLY`)//只存在内存中
211     */
212    def cache(): this.type = persist()
```

图 6-16

6.7.2 存储级别

默认的存储级别都是仅在内存中存储一份,Spark 的存储级别还有很多,存储级别在 object StorageLevel 中定义,如图 6-17 所示。

```
152  object StorageLevel {
153    val NONE = new StorageLevel(false, false, false, false)
154    val DISK_ONLY = new StorageLevel(true, false, false, false)
155    val DISK_ONLY_2 = new StorageLevel(true, false, false, false, 2)
156    val DISK_ONLY_3 = new StorageLevel(true, false, false, false, 3)
157    val MEMORY_ONLY = new StorageLevel(false, true, false, true)
158    val MEMORY_ONLY_2 = new StorageLevel(false, true, false, true, 2)
159    val MEMORY_ONLY_SER = new StorageLevel(false, true, false, false)
160    val MEMORY_ONLY_SER_2 = new StorageLevel(false, true, false, false, 2)
161    val MEMORY_AND_DISK = new StorageLevel(true, true, false, true)
162    val MEMORY_AND_DISK_2 = new StorageLevel(true, true, false, true, 2)
163    val MEMORY_AND_DISK_SER = new StorageLevel(true, true, false, false)
164    val MEMORY_AND_DISK_SER_2 = new StorageLevel(true, true, false, false, 2)
165    val OFF_HEAP = new StorageLevel(true, true, true, false, 1)
```

图 6-17

前面我们讲到的 persist 方法在调用时通过设置其参数进而实现缓存存储级别的自定义设置,即 persist 的参数可以指定持久化级别参数。

persist(MEMORY_ONLY) 表示将 RDD 以非序列化的 Java 对象存储于 JVM 中,如果内存不足,就要按照 LRU(Least Recently Used,最近最少使用)原则替换缓存中的内容。

persist(MEMORY_AND_DISK) 表示将 RDD 以非序列化的 Java 对象存储在 JVM 中,如果内存不足,超出的分区将会被存放在磁盘上。这里需要注意,并不是在内存和磁盘上都存放,而是优先使用内存,如果内存不足才会使用磁盘。

所以,使用 cache 方法时,会调用 persist(MEMORY_ONLY),即 cache() == persist(MEMORY_ONLY)。

可使用 unpersist 方法手动把持久化的 RDD 从缓存中移除。

Spark 提供了很多存储级别,在实际开发应用中应该如何选取持久化缓存的存储级别呢?实际上存储级别的选取就是内存与 CPU 之间的权衡,可参考以下内容。

- 如果 RDD 的数据可以很好地兼容默认存储级别（MEMORY_ONLY），那么优先使用它，这可以提高 CPU 运行速度。
- 尝试使用 MEMORY_ONLY_SER，且选择一种快速的序列化工具，也可以获得一种不错的效果。
- 一般情况下，不要把数据持久化到磁盘，除非计算是非常"昂贵"的或者计算过程会过滤掉大量数据，因为重新计算一个分区数据的速度可能要快于从磁盘读取一个分区数据的速度。
- 如果需要快速的失败恢复机制，则使用备份的存储级别，如 MEMORY_ONLY_2、MEMORY_AND_DISK_2 等。虽然所有的存储级别都可以通过重新计算丢失的数据实现容错，但是缓存机制使得大部分情况下应用程序无须中断，即数据丢失情况下，直接使用缓存数据，而不需要重新计算数据。
- 如果处于大内存或多应用的场景下，OFF_HEAP 可以带来以下好处：它允许 Spark Executor 共享 Tachyon 的内存数据，可很大程序上减少 JVM 垃圾回收带来的性能开销。Spark Executor 发生故障不会导致数据丢失。

最后，Spark 可以自己监测缓存空间的使用，并使用 LRU 算法移除旧的分区数据。我们也可以通过显式调用 unpersist 方法手动移除数据。

RDD 持久化缓存的存储级别如表 6-3 所示。

表 6-3

存储级别	说明
MEMORY_ONLY（默认）	将 RDD 以非序列化的 Java 对象存储在 JVM 中。如果没有足够的内存存储 RDD，则某些分区将不会被缓存，每次需要时都会重新计算。这是默认存储级别
MEMORY_AND_DISK（开发中可以使用这个）	将 RDD 以非序列化的 Java 对象存储在 JVM 中。如果数据在内存中放不下，则溢写到磁盘上，需要时从磁盘上读取
MEMORY_ONLY_SER（适用于 Java 和 Scala 语言）	将 RDD 以序列化的 Java 对象（每个分区一个字节数组）存储。这通常比非序列化对象更具空间效率，特别是在使用快速序列化的情况下，但是使用这种方式读取数据会消耗更多的 CPU 资源
MEMORY_AND_DISK_SER（适用于 Java 和 Scala 语言）	与 MEMORY_ONLY_SER 类似，但如果数据在内存中放不下，则溢写到磁盘上，而不是每次都重新计算它们
DISK_ONLY	将 RDD 分区存储在磁盘上
MEMORY_ONLY_2、MEMORY_AND_DISK_2 等	与上面的存储级别类似，只不过将持久化数据存为两份，备份每个分区存储在两个集群节点上
OFF_HEAP（实验中）	与 MEMORY_ONLY_SER 类似，但将数据存储在堆外内存中（即不直接存储在 JVM 中），如 Tachyon Alluxio（它们均是开源的基于内存的分布式存储系统）等

6.7.3 小结

RDD 持久化缓存是为了提高后续操作的速度。存储级别有很多，默认只存在内存中，开发中可以使用 MEMORY_AND_DISK。只有执行 Action 操作的时候才会真正将 RDD 进行持久化缓存。如果实际开发中某一个 RDD 后续会被频繁使用，可以将该 RDD 进行持久化缓存。

6.8 RDD 容错机制 Checkpoint

持久化缓存可以把数据放在内存中，虽然是快速的，但也是不可靠的。RDD 的容错机制 Checkpoint（检查点）保证了即使缓存丢失也能让计算正确执行。把数据放在 HDFS 上，这就天然地借助了 HDFS 的高容错、高可靠的特性，从而最大程度地保障数据安全。

由于 RDD 的各个分区是相对独立的，因此只需要计算丢失的部分。在实际应用开发中，我们可以通过下面所演示的两行代码来完成 Checkpoint 机制的应用。

```
SparkContext.setCheckpointDir("目录")  //HDFS的目录
RDD.Checkpoint  //Checkpoint保存
```

6.8.1 代码演示

我们继续使用 WordCount 案例演示 Checkpoint 机制，代码如下。

```
sc.setCheckpointDir("hdfs://master:9000/Checkpoint_dir")
//设置Checkpoint目录，会立即在HDFS上创建一个空目录
val rdd1 = sc.textFile("hdfs://master:9000/wordcount/input/words.txt")
.flatMap(_.split(" ")).map((_,1)).reduceByKey(_+_)
rdd1.Checkpoint  //对rdd1进行Checkpoint保存
rdd1.collect
//只有执行Action操作时才会真正执行Checkpoint，如果后面要使用rdd1，就可以从Checkpoint中直接读取
```

我们可以通过 HDFS 提供的命令查看 Checkpoint 的结果，代码如下。

```
hdfs dfs -ls hdfs://master:9000/Checkpoint_dir   //查看RDD容错目录的内容
drwxr-xr-x   - hadoop supergroup          0 2021-08-31 23:05 hdfs://master:9000/Checkpoint_dir/0913c14f-c42a-4d20-8f6b-178160ac05ce
```

或者通过 Web 界面查看（在浏览器输入 http://master:50070），结果如图 6-18 所示。

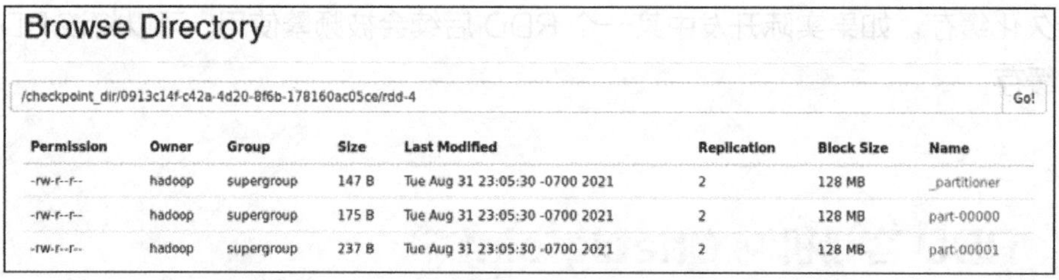
Browse Directory

/checkpoint_dir/0913c14f-c42a-4d20-8f6b-178160ac05ce/rdd-4   Go!

Permission	Owner	Group	Size	Last Modified	Replication	Block Size	Name
-rw-r--r--	hadoop	supergroup	147 B	Tue Aug 31 23:05:30 -0700 2021	2	128 MB	_partitioner
-rw-r--r--	hadoop	supergroup	175 B	Tue Aug 31 23:05:30 -0700 2021	2	128 MB	part-00000
-rw-r--r--	hadoop	supergroup	237 B	Tue Aug 31 23:05:30 -0700 2021	2	128 MB	part-00001

图 6-18

6.8.2 容错机制 Checkpoint 详解

分布式计算中难免因为网络、存储等原因出现计算失败的情况，RDD 中的 Lineage（血统、依赖关系）信息常供任务失败后重计算使用。为了防止计算失败后重新开始计算造成的大量开销，RDD 会计算过程的信息，作业失败后从 Checkpoint 重新计算即可。

所以，在开发中应保证数据的安全性及读取效率，可以对频繁使用且重要的数据先做持久化缓存，再做 Checkpoint 操作，其中我们需要明白持久化缓存和 Checkpoint 的区别。

首先，从存储位置上说，persist 和 cache 方法只能将 RDD 数据保存在本地磁盘和内存（或者堆外内存）；而 Checkpoint 容错机制可以将 RDD 数据保存到 HDFS 这类可靠的存储上。

其次，使用 persist 和 cache 方法，RDD 在程序结束后会被清除；而使用 Checkpoint，RDD 在程序结束后依然存在，不会被删除。

最后，persist 和 cache 方法不会丢掉 RDD 间的依赖链，即依赖关系，因为这种缓存是不可靠的，如果出现了一些错误（如宕机），需要通过回溯依赖链重新计算；而 Checkpoint 会斩断依赖链，因为 Checkpoint 会把结果保存在 HDFS 这类分布式文件存储系统中，更加安全、可靠，一般不需要回溯依赖链。

RDD 的 Lineage 会记录 RDD 的元数据信息和转换行为，当该 RDD 部分分区数据丢失时，它可以根据这些信息来重新计算和恢复丢失的数据分区，如图 6-19 所示。

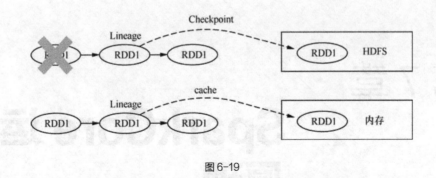

图 6-19

在进行故障恢复时，Spark 会对读取 Checkpoint 的开销和重新计算 RDD 分区的开销进行比较，从而自动选择最优的恢复策略。

6.9 本章总结

本章是 Spark 开发中的重中之重，介绍了 RDD 的相关核心概念和应用编程操作，包含 RDD 概念解析、RDD 应用编程、RDD 持久化缓存技术，以及 RDD 容错机制 Checkpoint。

6.10 本章习题

1. 理解 RDD 的核心概念，在此基础上通过代码实践 RDD 应用开发。

2. 总结 RDD 持久化缓存技术的应用。

第 7 章 SparkCore 运行原理

7.1 RDD 依赖关系

RDD 的依赖关系分为两种，即窄依赖（Narrow Dependency）与宽依赖（Wide Dependency，源码中称为 Shuffle Dependency），如图 7-1 所示。RDD 的依赖有两个重要作用：一是用来解决数据容错，二是用来划分阶段，即任务集合（TaskSet，后面详细介绍）。下面我们对这两种依赖关系的含义进行说明。

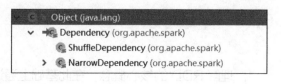

图 7-1

7.1.1 窄依赖与宽依赖

窄依赖：每个父 RDD 的一个分区最多被子 RDD 的一个分区所使用（1:1 或 n:1）。map、filter、union 等操作会产生窄依赖。

宽依赖：一个父 RDD 的分区会被多个子 RDD 的分区所使用。groupByKey、reduceByKey、sortByKey 等操作会产生宽依赖。窄依赖与宽依赖的业务模型如图 7-2 所示。

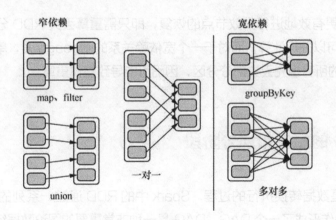

图 7-2

7.1.2 对比窄依赖与宽依赖

相比宽依赖，窄依赖对优化很有利，主要基于以下两点。第一，宽依赖对应 shuffle 操作，需要在运行过程中将同一个父 RDD 的分区传入不同的子 RDD 分区，中间可能涉及多个节点之间的数据传输；而窄依赖的每个父 RDD 的分区只会传入一个子 RDD 分区中，通常可以在一个节点内完成转换。

第二，当 RDD 分区丢失（某个节点故障）时，Spark 会对数据进行重算。对于窄依赖，由于父 RDD 的一个分区只对应一个子 RDD 分区，这样只需要重算和子 RDD 分区对应的父 RDD 分区即可，因此这个重算对数据的利用率是 100%。对于宽依赖，重算的父 RDD 分区对应多个子 RDD 分区，这样实际上父 RDD 中只有一部分数据是被用于恢复这个丢失的子 RDD 分区的，另一部分数据对应子 RDD 的其他未丢失分区，这就造成了多余的计算。而且，宽依赖中子 RDD 分区通常来自多个父 RDD 分区，极端情况下，所有的父 RDD 分区都要进行重算。

重算的效用不仅在于算多少，还在于有多少是冗余的计算。如图 7-3 所示，若 b1 分区丢失，则需要重算 a1、a2 和 a3，这就产生了冗余的计算（a1、a2、a3 中对应 b2 的数据）。

窄依赖允许在一个集群节点上以流水线的方式（Pipeline）计算所有父分区。例如，逐个元素执行 map 操作，然后执行 filter 操作。而宽依赖则需要首先计算好所有父分区数据，然后在节点之间进行 shuffle 操作。

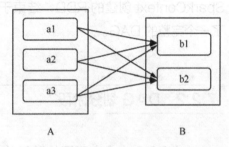

图 7-3

窄依赖能够更有效地进行失效节点的恢复，即只需重算丢失 RDD 分区的父分区，而且不同节点之间可以并行计算；而对于一个宽依赖关系的 Lineage 图，单个节点失效可能导致这个 RDD 的所有祖先丢失部分分区，因而可能导致整体重算。

7.2 DAG 的生成和划分阶段

DAG 指的是数据转换执行的过程。Spark 中的 RDD 通过一系列的 Transformation 操作和 Action 操作形成了一个 DAG。DAG 是一种非常重要的图论数据结构。从任何一个顶点出发经过若干条路径都无法回到最初的顶点，我们把这个图叫作 DAG，如图 7-4 所示。

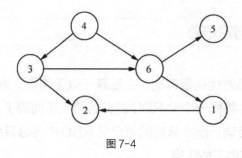

图 7-4

7.2.1 DAG 的生成

数据处理有方向，无闭环（其实就是 RDD 执行的流程）。原始的 RDD 通过一系列的 Transformation 操作形成了 DAG，任务执行时，可以按照 DAG 的描述，执行真正的计算（数据被操作的一个过程）。

一个 Spark 应用程序中可以有一到多个 DAG，这取决于触发了多少次 Action 操作。因为每触发一次 Action 操作就会产生一个 DAG，所以 DAG 的边界开始于通过 SparkContext 创建的 RDD，结束于触发 Action 操作，一旦触发 Action 操作，就形成了一个完整的 DAG。

7.2.2 DAG 划分阶段

一个 DAG 中会有不同的阶段（Stage），划分阶段的依据就是 RDD 依赖中的宽依赖。一个阶段中可以有多个任务，一个分区对应一个任务。Spark 会根据 shuffle/

宽依赖使用回溯算法从后向前遍历，遇到宽依赖就断开，遇到窄依赖就把当前的 RDD 加入阶段中，对于窄依赖，分区的转换处理在阶段中完成计算，不划分（将窄依赖尽量放在同一个阶段中，可以实现流水线计算）；对于宽依赖，由于有 shuffle 的存在，只能在父 RDD 处理完成后，才能开始接下来的计算，也就是说需要划分阶段，如图 7-5 所示。

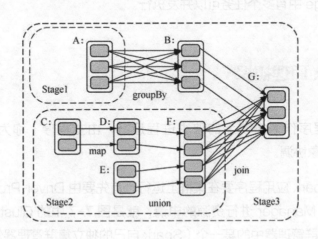

图 7-5

划分阶段主要是为了并行计算。对于一个复杂的业务逻辑，如果有 shuffle，那么意味着前面阶段产生结果后，才能执行下一个阶段，即下一个阶段的计算要依赖上一个阶段的数据。那么我们按照 shuffle 进行划分（也就是按照宽依赖执行划分），就可以将一个 DAG 划分成多个阶段，如图 7-5 中所划分出来的 Stage1、Stage2 和 Stage3。在同一个阶段中，会有多个算子操作，可以形成一个流水线，流水线内的多个平行的分区可以并发执行，如图 7-6 所示。

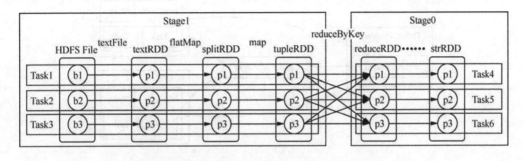

图 7-6

7.2.3 小结

Spark 会根据 shuffle/ 宽依赖使用回溯算法来对 DAG 进行阶段划分，从后往前，遇到宽依赖就断开，遇到窄依赖就把当前的 RDD 加入当前的阶段。每个阶段被称为一个 Stage，每个 Stage 中有多个任务可以并发执行。

7.3 Spark 原理初探

Spark 应用程序作为集群上的独立进程运行，由主程序（称为驱动程序）中的 SparkContext 对象协调。

具体来说，Spark 应用程序要在集群上运行，首先要由 Driver Program 连接到某一个具体的 Cluster Manager 进行资源的申请，就是图 7-7 中的 Cluster Manager 所代表的多种类型的集群管理器中的某一个（Spark 自己的独立集群管理器包括 Standalone、Mesos、YARN 或 Kubernetes 等平台），这些集群管理器可以跨应用程序分配资源。等连接 Cluster Manager 成功后，Spark 应用程序将会在该集群管理器所控制的集群中的各个 Worker 节点上申请计算资源以便获取 Executor，接收到计算任务的各个 Worker 节点将会为应用程序运行计算和存储数据的进程。接下来，SparkContext 将应用程序代码（由传递给 SparkContext 的 JAR 或 Python 文件定义）发送给 Executor。最后，SparkContext 将 Task 发送给 Executor 运行，如图 7-7 所示。

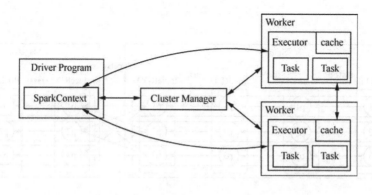

图 7-7

关于此体系结构，需要注意以下几点。

Spark 的每个应用程序都有自己的 Executor 进程，这些进程在整个应用程序运行期间保持不变，Executor 进程中会开辟多个线程同时运行多个任务。这样做的好处是在调度端（每个驱动程序调度自己的任务）和执行器端（不同 JVM 中运行的不同 Spark 应用程序的任务）将多个 Spark 应用程序彼此隔离。但是，这也意味着如果不将数据写入外部存储系统，就无法在不同的 Spark 应用程序（SparkContext 实例）之间共享数据。

Spark 中底层集群管理器具体是谁是不可知的，只要它能够从集群管理器中获取 Executor 进程。所以，Spark 不同应用程序就可以在不同类型的集群管理器上运行，例如在 Mesos/YARN 上运行。

驱动程序必须在整个生命周期内侦听并接收来自其 Executor 的传入连接，因此，驱动程序必须可以从工作节点进行网络寻址。

因为驱动程序在集群上调度任务，所以它应该在工作节点附近运行，最好在同一局域网上运行。如果你想远程向集群发送请求，最好打开一个 RPC 到驱动程序，让它从附近提交操作，而不是在远离工作节点的地方运行驱动程序。

7.3.1　Spark 相关的应用概念

Spark 应用程序在集群上独立运行，其中涉及很多相关的应用概念，下面对其进行详细说明。

1. Application

Application 指的是用户编写的 Spark 应用程序 / 代码，包含 Driver 功能代码和分布在集群中的多个节点上运行的 Executor 代码。一个 Application 由一个 Driver 和若干个作业（Job）构成，一个作业由多个阶段（Stage）构成，一个阶段由多个没有 shuffle 关系的任务（Task）组成，如图 7-8 所示。当执行一个 Application 时，Driver 会向集群管理器申请资源，启动 Executor，并向 Executor 发送应用程序代码和文件，然后在 Executor 上执行任务，运行结束后，执行结果会返回给 Driver，或者写到 HDFS 或者其他数据库中。

2. Driver

Spark 中的 Driver 负责运行上述 Application 的 main 方法并且创建 SparkContext，SparkContext 负责和 Cluster Manager 通信，进行资源的申请、任务的分配和监控等。

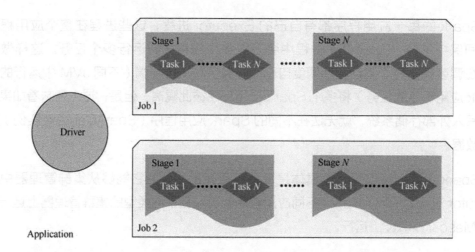

图 7-8

3. Cluster Manager

Cluster Manager 负责在集群上获取资源的外部服务，Standalone 模式下由 Master 负责，YARN 模式下由 Resource Manager 负责。

4. Executor

Executor 是运行在 Worker 节点上的进程，负责运行任务，并为应用程序存储数据，是执行分区计算任务的进程。

5. RDD

RDD 是分布式内存的一个抽象概念。

6. DAG

DAG 反映 RDD 之间的依赖关系和执行流程。

7. Job

程序，按照 DAG 执行就是一个程序。

8. Stage

阶段是作业的基本调度单位，同一个阶段中的任务可以并行执行，多个任务组成 TaskSet 集合。

9. Task

任务是运行在 Executor 上的工作单元，一个任务计算一个分区，包括流水线上的一系列操作。

7.3.2 Spark 基本流程概述

当一个 Spark 应用程序被提交时，首先需要为这个 Spark 应用程序构建基本的运行环境，即由 Driver 创建一个 SparkContext。

SparkContext 向资源管理器注册并申请运行 Executor 资源。

资源管理器为 Executor 分配资源并启动 Executor 进程，Executor 运行情况将随着心跳发送到资源管理器上。

SparkContext 根据 RDD 的依赖关系构建 DAG，并提交给 DAG 调度器（DAGScheduler）进行解析，划分成阶段，并把该阶段中的任务组成的 TaskSet 发送给任务调度器（TaskScheduler）。

任务调度器将任务发送给 Executor 运行，同时 SparkContext 将应用程序代码发送给 Executor。

Executor 将任务丢入线程池中运行，把运行结果反馈给任务调度器，然后反馈给 DAG 调度器，运行完毕后写入数据并释放所有资源。

7.3.3 流程图解

Spark 作业流程图如图 7-9 所示。

① 为应用程序构建基本的运行环境，即由 Driver 创建一个 SparkContext，进行资源的申请、任务的分配和监控等。

② 资源管理器为 Executor 分配资源，并启动 Executor 进程。

③ SparkContext 根据 RDD 的依赖关系构建 DAG，将 DAG 提交给 DAG 调度器进行解析，划分成阶段，然后把一个个任务集合 TaskSet 提交给底层任务调度器处理。SparkContext 将任务发送给 Executor 运行，并提供应用程序代码。

④ 任务在 Executor 上运行，把运行结果反馈给 SparkContext，运行完毕后写入数

据并释放所有资源。

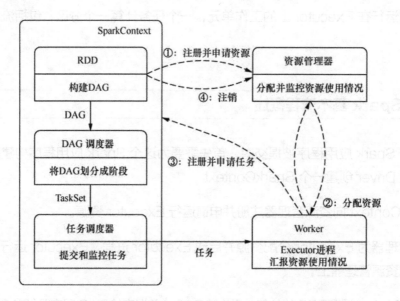

图 7-9

7.3.4 RDD 在 Spark 中的运行过程

RDD 在 Spark 中的运行过程分为 4 个阶段,如图 7-10 所示。

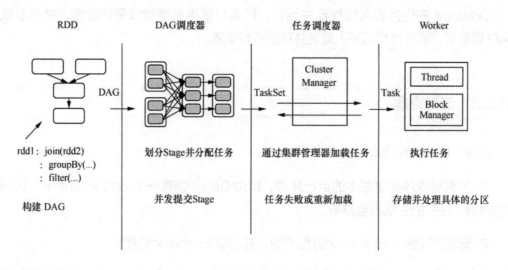

图 7-10

① 创建 RDD。

② SparkContext 负责计算 RDD 之间的依赖关系,构建 DAG。

③ DAG 调度器负责对 DAG 进行解析，划分成多个阶段，每个阶段中包含多个任务，每个任务会被任务调度器分发给各个 Worker 节点上的 Executor 去执行。

④ Spark 集群资源管理器中的 SchedulerBackend 定义了许多与 Executor 事件相关的处理，包括：新的 Executor 注册进来的时候记录 Executor 的信息；增加全局的资源量（核数）；Executor 更新状态。若任务完成后，Spark 集群资源管理器会回收 core 以及停止 executor、remove executor 等事件。

其中，DAG 调度器最主要的职责是将用户作业所形成的 DAG 划分成若干个阶段；任务调度器负责管理任务的分配及状态。在分配任务时，任务调度器会将 DAG 调度器提交的 TaskSet 进行优先级排序。这个排序算法目前有两种：FIFO 和 FAIR。得到这个待运行的任务后，接下来就是要把 Spark 集群资源管理器 SchedulerBackend 交过来的 Worker 资源信息合理分配给这些任务。

Executor 是真正执行任务的进程，本身拥有若干 CPU 和内存，可以执行以线程为单位的计算任务，它是资源管理系统能够给予的最小单位。

7.3.5 小结

Spark 运行架构的每个 Application 都有自己专属的 Executor 进程，并且该进程在 Application 运行期间一直驻留。Executor 进程以多线程的方式运行任务。Spark 运行过程与资源管理器无关，只要能够获取 Executor 进程并保持通信即可。

7.4 RDD 累加器和广播变量

默认情况下，当 Spark 应用程序在集群的多个不同节点的多个任务上并行运行一个函数时，它会把函数中涉及的每个变量，在每个任务上生成一个副本。但有时候需要在多个任务之间共享变量，或者在任务和 Driver 之间共享变量。为了满足这种需求，Spark 提供了两种类型的变量。

- 累加器（Accumulator），支持在所有不同节点之间进行累加计算，比如计数或者求和。
- 广播变量（Broadcast Variable），把变量在所有节点的内存之间进行共享，在每个机器上缓存一个只读的变量，而不是为机器上的每个任务都生成一个副本。

7.4.1 累加器

累加器是仅仅被相关操作累加的变量，通常可以被用来实现计数器（counter）和求和（sum）。Spark 支持数值型（numeric）的累加器，程序开发人员可以编写支持新类型的代码。

一个数值型的累加器，可以通过调用 SparkContext.longAccumulator 或者 SparkContext.doubleAccumulator 来创建。

运行在集群中的任务，就可以使用 add 方法把数值累加到累加器上。但是，这些任务只能做累加操作，不能读取累加器的值，只有 Driver 可以使用 value 方法来读取累加器的值。

下面就来实现一个累加器。我们先来看一下，如果不使用 Spark 提供的累加器是什么效果。

```
var counter = 0
val data = Seq(1,2,3)
data.foreach(x => counter += x)
println("Counter value" + counter)
```

运行结果如下。

```
Counter value: 6
```

如果我们将 data 转化为 RDD，再来重新计算计数器。

```
var counter = 0
val data = Seq(1,2,3)
var rdd = sc.parallelize(data)
rdd.foreach(x => counter += x)
println("Counter value:" + counter)
```

运行结果如下。

```
Counter value: 0
```

当 data 转化为 RDD 运行时，结果 Counter value 为 0。这是因为 foreach 函数加载给 Worker 中的 Executor 执行，用到了 counter 变量，而 counter 变量是在 Driver 端定义的，在传递给 Executor 的时候，各个 Executor 都有一个 counter。最后各个 Executor 将各自的 x 加到自己的 counter 上面，和 Driver 端的 counter 没有关系，所以在 Driver 端输出 counter 的值为 0。

如何解决这个问题？使用 Spark 提供的累加器就可以了。

通常在向 Spark 传递函数时，比如使用 map 函数或者用 filter 传条件时，可以使用驱动程序中定义的变量，但是集群中运行的每个任务都会得到这些变量的一个新的副本，更新这些副本的值也不会影响驱动器中的对应变量。这时使用累加器就可以实现我们想要的效果。

```
//整数累加器
val xx: Accumulator[Int] = SparkContext.accumulator(0)
//长整数累加器
val xx1: Accumulator[Int] = SparkContext.longAccumulator(0)
//Double累加器
val xx2: Accumulator[Int] = SparkContext.doubleAccumulator(0)
```

下面通过一段完整的代码来演示 Spark 提供的累加器的功能。

```
package cn.itcast.core
import org.apache.spark.rdd.RDD
import org.apache.spark.{Accumulator, SparkConf, SparkContext}

object MyAccumulatorTest {
  def main(args: Array[String]): Unit = {
    val conf: SparkConf = new SparkConf().setAppName("wc").setMaster("local[*]")
    val sc: SparkContext = new SparkContext(conf)
    sc.setLogLevel("WARN")
    //使用Scala编程集合完成累加
    var counter1: Int = 0;
    var data = Seq(1,2,3)
    data.foreach(x => counter1 += x )
    println(counter1)//6
    println("++++++++++++++++++++++++")
    //使用RDD进行累加
    var counter2: Int = 0;
    val dataRDD: RDD[Int] = sc.parallelize(data)  //分布式集合的[1,2,3]
    dataRDD.foreach(x => counter2 += x)
    println(counter2)  //0
    //注意：上面的RDD操作运行结果是0
    //因为foreach中的函数是传递给Worker中的Executor执行，用到了counter2变量
    //counter2是在Driver端定义的，在传递给Executor时，各个Executor都有一个counter2
    //最后各个Executor将各自的x加到自己的counter2上面，和Driver端的counter2没有关系
    //这个问题需要解决！不能因为使用了Spark连累加都做不了
    //如何解决？使用Spark提供的累加器
    val counter3: Accumulator[Int] = sc.accumulator(0)
```

```
    dataRDD.foreach(x => counter3 += x)
    println(counter3) //6
}
}
```

7.4.2 广播变量

广播变量允许在每台机器上缓存一个只读的变量,而不是为机器上的每个任务都生成一个副本。

如果不使用广播变量,就会在每个任务上生成只读变量的一个副本。如图 7-11 所示,有两个 Worker 节点,每个 Worke 节点上有两个任务,所以就会产生 4 个只读变量的副本。

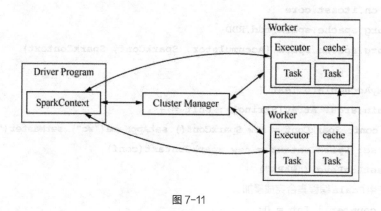

图 7-11

如果使用广播变量,就会在每台机器上生成只读变量的一个副本。如图 7-12 所示,有两个 Worker 节点,就会产生两个只读变量的副本,并且会在 Spark Context 上实现共享变量。

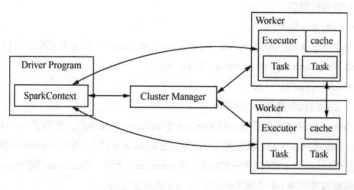

图 7-12

Spark 的 Action 操作会跨越多个阶段,对于每个阶段内的所有任务所需要的公共数据，Spark 都会自动进行广播。

可以通过调用 SparkContext.broadcast(v) 来从一个普通变量 v 中创建一个广播变量。这个广播变量就是对普通变量 v 的一个包装器，通过调用 value 方法就可以获得这个广播变量的值，代码如下。

```
val broadcastVar = sc.broadcast(Array(1, 2, 3))
broadcastVar.value
```

这个广播变量被创建以后，在集群中的任何函数中，都可以使用广播变量的值，这样就不会把 v 重复分发到这些节点上。此外，一旦广播变量被创建，普通变量 v 的值就不能再变化，从而确保所有节点获得的这个广播变量的值都相同。

下面我们通过一段代码来演示广播变量的实际应用。

```
package cn.itcast.core

import org.apache.spark.broadcast.Broadcast
import org.apache.spark.rdd.RDD
import org.apache.spark.{SparkConf, SparkContext}

object MyBroadcastVariables {
  def main(args: Array[String]): Unit = {
    val conf: SparkConf = new SparkConf().setAppName("wc").setMaster("local[*]")
    val sc: SparkContext = new SparkContext(conf)
    sc.setLogLevel("WARN")

    //不使用广播变量
    val kvFruit: RDD[(Int, String)] = sc.parallelize(List((1,"apple"),(2,"orange"),(3,"banana"),(4,"grape")))
    val fruitMap: collection.Map[Int, String] =kvFruit.collectAsMap
    /*scala.collection.Map[Int,String] = Map(2 -> orange, 4 -> grape, 1 -> apple, 3 -> banana)*/
    val fruitIds: RDD[Int] = sc.parallelize(List(2,4,1,3))
    //根据水果编号取水果名称
    val fruitNames: RDD[String] = fruitIds.map(x=>fruitMap(x))
    fruitNames.foreach(println)
    //注意：以上代码看似没有问题，但是考虑到数据量如果较大，且任务数较多，
    //那么会导致各个任务共用的fruitMap被多次传输
    //应该减少fruitMap的传输，一台机器上一个，被该台机器中的任务共用即可
    //如何做到？使用广播变量
    println("====================")
```

```
    val BroadcastFruitMap: Broadcast[collection.Map[Int, String]] = sc.broad-
cast(fruitMap)
    val fruitNames2: RDD[String] = fruitIds.map(x=>BroadcastFruitMap.value(x))
    fruitNames2.foreach(println)
  }
}
```

7.5 RDD 的数据源

RDD 的数据源有多种，可以是普通文本文件、HDFS、SequenceFile、对象文件、HBase、JDBC 等，如图 7-13 所示。其中，普通文本文件、HDFS 和 JDBC 数据源要重点掌握。

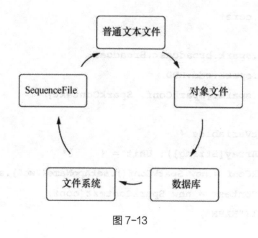

图 7-13

7.5.1 普通文本文件

加载普通文本文件数据的方法，代码如下。

```
sc.textFile("./dir/*.txt")
```

如果传递目录，则将目录下的所有文件读取为 RDD。文件路径支持通配符。但是这样对于大量的小文件读取效率并不高，应该使用 wholeTextFiles，代码如下。

```
def wholeTextFiles(path: String, minPartitions: Int = defaultMinParti-
tions): RDD[(String, String)])
```

返回值是 RDD[(String, String)]，其中 key 是文件的名称，value 是文件的内容。

① 从本地文件系统 D:/java/Spark/day11_spark/sample1.dat 获取 RDD 数据源。

val file: RDD[String]=sc.textFile("D:/java/Spark/day11_spark/sample1.dat")

② 读取小文件。

```
println("读取小文件")
val filesRDD: RDD[(String, String)] = sc.wholeTextFiles("D:\\data\\spark\\files", minPartitions = 3)
val linesRDD: RDD[String] = filesRDD.flatMap(_._2.split("\\r\\n"))
val wordsRDD: RDD[String] = linesRDD.flatMap(_.split(" "))
wordsRDD.map((_, 1)).reduceByKey(_ + _).collect().foreach(println)
```

7.5.2 Hadoop API

Spark 的整个生态系统与 Hadoop 是完全兼容的，所以 Hadoop 所支持的文件类型或者数据库类型，Spark 同样支持。HadoopRDD、newAPIHadoopRDD、saveAsHadoopFile、saveAsNewAPIHadoopFile 是底层 API，其他的 API 都是为了方便 Spark 应用程序开发者而设置的，是这两个接口的高效实现版本。

Spark 源码如图 7-14 所示。

SparkContext.scala

```
  */
def textFile(
    path: String,
    minPartitions: Int = defaultMinPartitions): RDD[String] = withScope {
  assertNotStopped()
  hadoopFile(path, classOf[TextInputFormat], classOf[LongWritable], classOf[Text],
    minPartitions).map(pair => pair._2.toString).setName(path)
}
```

RDD.scala

```
  */
def saveAsTextFile(path: String): Unit = withScope {
  // https://issues.apache.org/jira/browse/SPARK-2075
  //
  // NullWritable is a `Comparable` in Hadoop 1.+, so the compiler cannot find an implicit
  // Ordering for it and will use the default `null`. However, it's a `Comparable[NullWritable]`
  // in Hadoop 2.+, so the compiler will call the implicit `Ordering.ordered` method to create an
  // Ordering for `NullWritable`. That's why the compiler will generate different anonymous
  // classes for `saveAsTextFile` in Hadoop 1.+ and Hadoop 2.+.
  //
  // Therefore, here we provide an explicit Ordering `null` to make sure the compiler generate
  // same bytecodes for `saveAsTextFile`.
  val nullWritableClassTag = implicitly[ClassTag[NullWritable]]
  val textClassTag = implicitly[ClassTag[Text]]
  val r = this.mapPartitions { iter =>
    val text = new Text()
    iter.map { x =>
      text.set(x.toString)
      (NullWritable.get(), text)
    }
  }
  RDD.rddToPairRDDFunctions(r)(nullWritableClassTag, textClassTag, ord = null)
    .saveAsHadoopFile[TextOutputFormat[NullWritable, Text]](path)
}
```

图 7-14

Hadoop API 应用案例代码如下。

```
println("Hadoop API")
//创建dataRDD
val dataRDD = sc.parallelize(Array((1,"hadoop"), (2,"hive"), (3,"spark")))
//将dataRDD写入HDFS中，key的类型为LongWritable、value的类型为Text
    dataRDD.saveAsNewAPIHadoopFile("hdfs://master:9000/spark_hadoop/",
        classOf[LongWritable],
        classOf[Text],
        classOf[TextOutputFormat[LongWritable, Text]])
//读取hdfs://master:9000/spark_hadoop/*大量的小文件
    val inputRDD: RDD[(LongWritable, Text)] = sc.newAPIHadoopFile(
        "hdfs://master:9000/spark_hadoop/*",
        classOf[TextInputFormat],
        classOf[LongWritable],
        classOf[Text],
        conf = sc.hadoopConfiguration
    )
    inputRDD.map(_._2.toString).foreach(println)
```

7.5.3 SequenceFile

SequenceFile 是 Hadoop 用来存储二进制形式的 key-value 而设计的一种平面文件（Flat File），如图 7-15 所示。

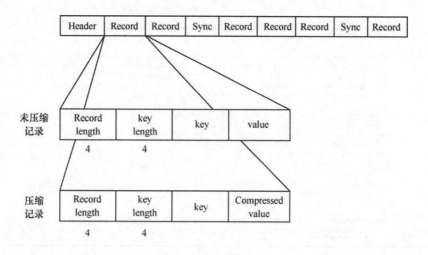

图 7-15

序列化文件案例代码如下。

读取命令：`sc.sequenceFile[ keyClass, valueClass](path)`
写入命令：`RDD.saveAsSequenceFile(path)`

```
println("SequenceFile")
//创建dataRDD
val dataRDD: RDD[(Int, String)] = sc.parallelize(List((2, "aa"), (3, "bb"), (4, "cc"), (5, "dd"), (6, "ee")))
//将dataRDD序列化并存储到目录D:\\data\\spark\\SequenceFile
dataRDD.saveAsSequenceFile("D:\\data\\spark\\SequenceFile")
//读取序列化文件，并输出
val sdata: RDD[(Int, String)] = sc.sequenceFile[Int, String]("D:\\data\\spark\\SequenceFile\\*")
    sdata.collect().foreach(println)
```

在序列化文件的操作过程中，要求 key 和 value 能够自动转变为 Writable 类型，如图 7-16 所示。

Scala类型	Java类型	Writable类型
Int	Integer	IntWritable
Long	Long	LongWritable
Float	Float	FloatWritable
Double	Double	DoubleWritable
Boolean	Boolean	BooleanWritable
Array[Byte]	byte[]	BytesWritable
String	String	Text
Array[T]	T[]	ArrayWritable<TW>
List[T]	List<T>	ArrayWritable<TW>
Map[A,B]	Map<A，B>	MapWritable<AW,BW>

图 7-16

7.5.4 对象文件

对象文件是将对象序列化后保存的文件。

读取使用 sc.objectFile[k,v](path)，其中 [k,v] 用来指定对象文件内容的数据类型，这是反序列化时对对象类型的指定，因为该方法用来读取序列化的对象文件；sc 即 SparkContext。

写入使用 RDD.saveAsObjectFile()。

操作对象文件案例代码如下。

```
println("ObjectFile")
//创建dataRDD
    val dataRDD = sc.parallelize(List((2, "aa"), (3, "bb"), (4, "cc"), (5, "dd"),
(6, "ee")))
    //保存对象文件
    dataRDD3.saveAsObjectFile("D:\\data\\spark\\ObjectFile")
    //读取对象文件，注意读取时需要指定类型
    val objRDD = sc.objectFile[(Int, String)]("D:\\data\\spark\\ObjectFile\\*")
    objRDD.collect().foreach(println)
```

7.5.5 HBase

Spark 可以通过 Hadoop 输入格式访问 HBase，由 org.apache.hadoop.hbase.mapreduce.TableInputFormat 类实现，这个输入格式会返回 key-value 数据，其中 key 的类型为 org.apache.hadoop.hbase.io.ImmutableBytesWritable，而 value 的类型为 org.apache.hadoop.hbase.client.Result。

访问 HBase 案例代码如下。

```
package cn.stone.core
import org.apache.hadoop.hbase.client.{HBaseAdmin, Put, Result}
import org.apache.hadoop.hbase.io.ImmutableBytesWritable
import org.apache.hadoop.hbase.{HBaseConfiguration, HColumnDescriptor, HTableDescriptor, TableName}
import org.apache.hadoop.hbase.mapred.TableOutputFormat
import org.apache.hadoop.hbase.mapreduce.TableInputFormat
import org.apache.hadoop.hbase.util.Bytes
import org.apache.hadoop.mapred.JobConf
import org.apache.spark.rdd.RDD
import org.apache.spark.{SparkConf, SparkContext}

object DataSourceTest2 {
  def main(args: Array[String]): Unit = {
    //配置信息
    val config = new SparkConf().setAppName("DataSourceTest").setMaster("local[*]")
    val sc = new SparkContext(config)
```

```scala
    sc.setLogLevel("WARN")
    //创建访问HBase数据库的配置信息
    val conf = HBaseConfiguration.create()
    conf.set("hbase.zookeeper.quorum", "master:2181,slave:2181,slave1:2181")
    //获取表的名字信息对象
    val fruitTable = TableName.valueOf("fruit")
    //创建表的描述信息对象
    val tableDescr = new HTableDescriptor(fruitTable)
    //为表创建列族
    tableDescr.addFamily(new HColumnDescriptor("info".getBytes))
    //获取访问HBase数据库的对象,相当于连接信息
    val admin = new HBaseAdmin(conf)

//判断fruitTable是否存在,如果存在先进行删除
    if (admin.tableExists(fruitTable)) {
      admin.disableTable(fruitTable)
      admin.deleteTable(fruitTable)
    }
    //创建表
    admin.createTable(tableDescr)
    /*
定义convert函数,将字符串转换为类型为HBase的对象的key-value。
            输入:3个字符串。
            输出:key-value
            */
    def convert(triple: (String, String, String)) = {
      val put = new Put(Bytes.toBytes(triple._1))
      put.addImmutable(Bytes.toBytes("info"),
Bytes.toBytes("name"), Bytes.toBytes(triple._2))
      put.addImmutable(Bytes.toBytes("info"),
Bytes.toBytes("price"), Bytes.toBytes(triple._3))
      (new ImmutableBytesWritable, put)
    }
    //加载模拟数据
    val dataRDD: RDD[(String, String, String)] = sc.parallelize(List(("1","apple",
"11"), ("2","banana","12"), ("3","pear","13")))
    //调用convert函数,将数据转化为HBase对象的key-value
    val targetRDD: RDD[(ImmutableBytesWritable, Put)] = dataRDD.map(convert)
    //创建作业的配置信息对象
    val jobConf = new JobConf(conf)
//设置作业运行的输出格式为TableOutputFormat
```

```
jobConf.setOutputFormat(classOf[TableOutputFormat])
//设置输出表的名字为 fruit
jobConf.set(TableOutputFormat.OUTPUT_TABLE, "fruit")
//向 HBase 数据库中写入数据
targetRDD.saveAsHadoopDataset(jobConf)
println("写入数据成功")

//从 HBase 数据库中读取数据
//设置输入数据的格式为 TableInputFormat 及输入表的名字为 fruit
conf.set(TableInputFormat.INPUT_TABLE, "fruit")
//获取查询表结果信息
 val hbaseRDD: RDD[(ImmutableBytesWritable, Result)] = sc.newAPIHadoopRDD(conf, classOf[TableInputFormat],
      classOf[org.apache.hadoop.hbase.io.ImmutableBytesWritable],
      classOf[org.apache.hadoop.hbase.client.Result])
//获取数据的总条数
val count: Long = hbaseRDD.count()
//输出数据的总条数
println("hBaseRDD RDD Count:"+ count)
//遍历查询结果信息
hbaseRDD.foreach {
   case (_, result) =>
     //获取 key 字段的值
     val key = Bytes.toString(result.getRow)
     //获取 name 字段的值
     val name = Bytes.toString(result.getValue("info".getBytes, "name".getBytes))
     //获取 color 字段的值
     val color = Bytes.toString(result.getValue("info".getBytes, "price".getBytes))
     //输出 key、name 和 color 这 3 个字段的值信息
     println("Row key:" + key + " Name:" + name + " Color:" + color)
  }
   sc.stop()
  }
 }
```

7.5.6　JDBC

Spark 支持通过 JDBC 访问关系数据库。需要使用 JdbcRDD API。首先来认识一下 JdbcRDD API，如图 7-17 所示。

```
new JdbcRDD(sc: SparkContext, getConnection: () ⇒ Connection, sql: String,
    lowerBound: Long, upperBound: Long, numPartitions: Int, mapRow:
    (ResultSet) ⇒ T = JdbcRDD.resultSetToObjectArray)(implicit arg0:
    ClassTag[T])
```

getConnection	获取数据库的链接
sql	用于查询数据库的sql语句
	select title, author from books where ? <= id and id <= ?
lowerBound	第一个占位符的最小值
upperBound	第二个占位符的最小值
numPartitions	包含下限和上限的分区数量
mapRow	查询返回结果集，以数组的方式呈现

图 7-17

- getConnection：获取数据库的连接。
- sql：访问数据库的 SQL 语句。
- lowerBound：第一个占位符的最小值，例如 ID 值的最小范围。
- upperBound：第二个占位符的最大值，包括下限和上限。
- numPartitions：分区的数量。如果下限为 1，上限为 20，numPartitions 为 2，则查询将执行两次，一次使用 (1,10)，一次使用 (11,20)。
- mapRow：从结果集到所需结果类型的单行函数。这里只调用 getInt、getString 等，RDD 负责调用 next。默认情况下，将结果集映射到对象数组。

下面来演示使用 Spark 操作 JdbcRDD API 实现将数据存入 MySQL 并读取出来，代码如下。

```
package cn.stone.core
import java.sql.{Connection, DriverManager, PreparedStatement}
import org.apache.spark.rdd.{JdbcRDD, RDD}
import org.apache.spark.{SparkConf, SparkContext}
/**
 * 演示使用Spark操作JdbcRDD API实现将数据存入MySQL并读取出来
 */
object JDBCDataSourceTest {
  def main(args: Array[String]): Unit = {
    //1.创建SparkContext
    val config = new SparkConf()
.setAppName("JDBCDataSourceTest").setMaster("local[*]")
    val sc = new SparkContext(config)
```

```scala
      sc.setLogLevel("WARN")
    //2.插入数据
    val data: RDD[(String, Int)] = sc.parallelize(List(("jack", 18), ("tom", 19), ("rose", 20)))
    //调用foreachPartition针对每一个分区进行操作
    data.foreachPartition(saveToMySQL)

    //3.读取数据
    def getConn():Connection={
      DriverManager.getConnection("jdbc:mysql://localhost:3306/bigdata?characterEncoding=UTF-8", "root", "root")
    }
    val studentRDD: JdbcRDD[(Int, String, Int)] = new JdbcRDD(sc,
      getConn,
      "select * from t_student where id >= ? and id <= ? ",
      4,
      6,
      2,
      rs => {
        val id: Int = rs.getInt("id")
        val name: String = rs.getString("name")
        val age: Int = rs.getInt("age")
        (id, name, age)
      }
    )
    println(studentRDD.collect().toBuffer)
  }
  /*  在MySQL中创建t_student表信息
      CREATE TABLE 't_student' (
        'id' int(11) NOT NULL AUTO_INCREMENT,
        'name' varchar(255) DEFAULT NULL,
        'age' int(11) DEFAULT NULL,
        PRIMARY KEY ('id')
      ) ENGINE=InnoDB AUTO_INCREMENT=4 DEFAULT CHARSET=utf8;
  */

  def saveToMySQL(partitionData:Iterator[(String, Int)] ):Unit = {
    //将数据存入MySQL
    //获取MySQL数据库连接
    val conn: Connection = DriverManager.getConnection("jdbc:mysql://localhost:3306/bigdata?characterEncoding=UTF-8", "root", "root")
```

```
    partitionData.foreach(data=>{
       //将每一条数据存入MySQL
       val sql = "INSERT INTO 't_student' ('id', 'name', 'age') VALUES (NULL, ?, ?);"
       val ps: PreparedStatement = conn.prepareStatement(sql)
       ps.setString(1,data._1)
       ps.setInt(2,data._2)
       ps.execute()//preparedStatement.addBatch()
    })
//ps.executeBatch()
    conn.close()
  }
}
```

7.6 本章总结

本章继续围绕 SparkCore 编程技术进行详解,主要介绍了 Spark 应用程序在数据处理中的 DAG 生成、根据 RDD 宽依赖对阶段的划分、Spark 框架原理的初步探索,使读者掌握 Spark 基本的运行过程。最后介绍了 RDD 的多种数据源,包括普通文本文件数据、HDFS、HBase 以及 JDBC 数据源等。

7.7 本章习题

1. 画出 RDD 依赖关系图并进行阶段的划分。

2. 通过代码实践 RDD 的文件数据、HDFS、HBase 以及 JDBC 等数据的开发应用。

第 8 章

Spark SQL 结构化数据处理入门

Spark SQL 是 Spark 实时计算框架的一个子模块，用来对结构化数据进行分析与处理。与 SparkRDD 不同，Spark SQL 的众多接口为数据处理提供了数据结构化和计算本身演绎的更多信息。Spark SQL 提供了多种与其交互的方法：SQL 和 Dataset API。无论采用哪种方法，在计算结果时，将会使用相同的底层执行引擎，其与用于实现计算的 API 或语言无关。这种统一的底层数据处理方式，可以帮助数据分析人员轻松地在不同的 API 之间切换。

8.1 数据分析方式

在 Spark 中数据分析方式有两种：一种是前面已经学过的命令式；另一种是我们所熟悉的 SQL 式。

8.1.1 命令式

RDD 是命令式的，主要特征是通过一个算子，可以计算得到一个结果，通过结果再进行后续计算。

```
sc.textFile("...")
  .flatMap(_.split(" "))
```

```
.map((_, 1))
.reduceByKey(_ + _)
.collect()
```

命令式的数据分析方式有 3 个特点：一是数据操作粒度更细，能够控制数据的每一个处理环节；二是操作更明确、步骤更清晰、容易维护；三是支持半/非结构化数据的操作。总之，命令式的数据分析方式灵活，数据的范围也更广。

但命令式的数据分析方式也有很致命的缺点：第一，需要开发者具有一定的编程功底；第二，代码写起来比较麻烦，在实际开发中效率很低。

8.1.2 SQL 式

SQL 式是一种数据分析的标准，已经存在多年。随着数据规模的日渐增大，原有的分析技巧是否就过时了呢？答案显然是否定的，原有的分析技巧在既有的分析维度上依然有效。

那么数据分析人员如何快速适应大数据平台呢？去重新学一种脚本吗？直接用 Scala 或 Python 去编写 RDD 吗？显然这样的学习成本过大。数据分析人员希望"在大数据平台仍然直接使用 SQL 式的数据分析方式"。

SQL 式的数据分析方式的优点非常显著，语言简洁，语法简单，使用方式灵活等，比如下面这条 SQL 语句通过 3 个字段来查询年龄大于 30 岁的用户信息。

```
SELECT
    username,
    age,
    sex
FROM customers
WHERE age > 30
```

当然 SQL 式的数据分析方式的缺点也很明显，如果碰到 3 层嵌套的 SQL，维护起来就力不从心。想象一下，如果使用 SQL 来实现机器学习算法是很艰难的。

因此，SQL 式擅长数据分析和通过简单的语法表示查询，命令式操作适合过程式处理和算法性处理。

在 Spark 出现之前，一款工具通常只支持 SQL 式或命令式，进行结构化数据的查询和处理时，开发者被迫要使用多个工具来适应两种场景，并且多个工具配合起来比较复杂。例如，对于数据分析和通过简单语法表示的查询，需要使用 Hive 工具；对于过程式和算法性处理，需要编写 MapReduce 程序。而 Spark 的出现统一了两种数据处理范式。

8.2 Spark SQL 的发展

SQL 是数据分析领域一个非常重要的范式，Hive 的流行直接证明了 SQL 在大数据平台的设计迎合了技术市场的需求，所以 Spark 一直想要支持这种范式，而伴随着一些决策的失误，这个过程其实还是非常曲折的。

Spark SQL 的发展经历了 4 个阶段：一是需要 SQL on Hadoop；二是 Hive 的成功；三是 Shark 时期；四是 Spark SQL 蜕变完成，如图 8-1 所示。

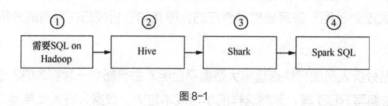

图 8-1

在这个过程中，Hive 实现了 SQL on Hadoop 的需求，底层的计算处理引擎用的是 MapReduce 分布式离线计算框架，从而简化了 MapReduce 的任务，开发人员不用再去写复杂、冗余的 MapReduce 程序，直接用 SQL 就可以实现数据分析和简单的查询。Hive 框架会自动将 SQL 语句转化为 MapReduce 程序并在 Hadoop 平台上执行。

但是，新的问题出现了，Hive 查询延迟比较高，执行一条 SQL 查询语句需要很长时间，其原因就是底层数据计算处理引擎使用了 MapReduce。

而 Shark 框架改写了 Hive 的物理执行计划，使用 Spark 代替 MapReduce 的物理引擎，使用列式内存存储方式。以上两点使得 Shark 的查询效率非常高。

为了实现与 Hive 兼容，Shark 在 HQL 方面重用了 Hive 中 HQL 的解析、逻辑执行计划翻译、执行计划优化等逻辑，可以认为仅将物理执行计划从 MapReduce 作业替换成了 Spark 作业，通过 Hive 中 HQL 的解析，把 HQL 翻译成 Spark 上的 RDD 操作。

因此，Shark 的出现使得 SQL on Hadoop 的性能与 Hive 相比有了显著的提高。

但是，Shark 的设计导致了两个问题：第一，Shark 执行计划的生成严重依赖 Hive，想要增加新的优化策略非常困难；第二，Spark 是线程级别的并行，而 Hive 的底层是 MapReduce，因此是进程级别的并行。Spark 在兼容 Hive 的实现上存在线程安全问题，导致 Shark 不得不使用另外一套独立维护的打了补丁的 Hive 源码分支，但该分支无法合并进主线。

2014 年 6 月 1 日，Shark 项目和 Spark SQL 项目主持人宣布停止对 Shark 项目的开发，团队将所有资源放在 Spark SQL 项目上。至此，Shark 的发展画上了句号，但也因此发展出两条线路：Spark SQL 和 Hive on Spark。

Spark SQL 作为 Spark 生态的一员继续发展，不再受限于 Hive，只是兼容 Hive。Hive on Spark 是一个 Hive 的发展计划，该计划将 Spark 作为 Hive 的底层引擎之一。因此 Hive 可以采用 MapReduce、Tez、Spark 等底层计算引擎。

Spark SQL 执行计划和优化交给了优化器 Catalyst，其内建了一套简单的 SQL 解析器，可以不使用 HiveQL，还可引入 DataFrame 这样的 DSL API，完全不依赖任何 Hive 组件。

Spark 1.6 引入了一个 API，叫作 Dataset。Dataset 统一和结合了 SQL 的访问和命令式 API 的使用，这是一个"划时代"的进步。在 Dataset 中可以轻松使用 SQL 查询并且筛选数据，然后使用命令式 API 进行探索式分析。

8.3 数据分类和 Spark SQL 适用场景

本节主要讲解结构化数据、半结构化数据和非结构化数据。

8.3.1 结构化数据

结构化数据也称作行数据，是由二维表结构来表达逻辑和实现的数据，其严格地遵循数据格式与长度规范，主要通过关系数据库进行存储和管理。

结构化数据一般有固定的 Schema（约束）。例如在用户表中，name 字段是 String 类型的，那么每一条数据的 name 字段值都可以当成字符串来使用。

与结构化数据相对的是不适合由数据库二维表来表现的非结构化数据，包括所有格式的办公文档、XML 文件、HTML 文件、各类报表、图片和音频 / 视频信息等。

8.3.2 半结构化数据

半结构化数据一般没有固定的 Schema，但是数据本身是有结构的。比如一个用户信息的 JSON 文件，第一条数据的 phone_num 有可能是数字，第二条数据的 phone_num 虽说应该也是数字，但是如果指定为字符串，也是可以的，因为没有指定固定的 Schema 信息，没有显式的强制约束。

虽说半结构化数据没有显式指定 Schema，也没有约束，但是半结构化数据本身是有隐式结构的，也就是数据自身可以描述自身。如果 JSON 文件中的某一条数据是有字段这个概念的，每个字段也有类型的概念，那么 JSON 文件是可以描述自身的，也就是数据本身携带元数据信息。

一个 JSON 文件的内容如下。

```
{
    "firstName": "John",
    "lastName": "Smith",
    "age": 25,
    "phoneNumber":
    [
        {
          "type": "home",
          "number": 212 555-1234
        },
        {
          "type": "fax",
          "number": "646 555-4567"
        }
    ]
}
{
    "firstName": "Johnz",
    "lastName": "Smithz",
    "hobby": "java",
    "phoneNumber":
    [
        {
          "type": "home",
```

```
            "number": "212 555-1234"
        },
        {
            "type": "fax",
            "number": "646 555-4568"
        }
    ]
}
```

8.3.3 非结构化数据

非结构化数据是数据结构不规则或不完整、没有预定义的数据模型、不方便用数据库二维表来表现的数据，包括所有格式的办公文档、文本、XML 文件、HTML 文件、各类报表、图片和音频 / 视频信息等。

非结构化数据的格式非常多样，标准也是多样性的，而且在技术上非结构化数据比结构化数据更难标准化和理解。所以存储、检索、发布以及利用非结构化数据需要更加智能化的信息技术，比如海量存储、智能检索、知识挖掘、内容保护、信息的增值开发利用等。

以上我们介绍了结构化数据、半结构化数据和非结构化数据，针对这 3 种格式的数据进行对比和总结，如表 8-1 所示。

表 8-1

数据分类	定义	举例
结构化数据	有固定的 Schema	关系数据库的表
半结构化数据	没有固定的 Schema，但是有结构	一些有结构的文件，例如 JSON 文件
非结构化数据	没有固定的 Schema，也没有结构	图片、音频等

思考一个问题，我们学习的 Spark 框架处理的是什么样的数据呢？首先，RDD 主要用于处理非结构化数据、半结构化数据和结构化数据。其次，Spark SQL 主要用于处理结构化数据和较为规范的半结构化数据。

那么 Spark SQL 相较于 RDD 的优势在哪里？第一，Spark SQL 提供了更好的外部数据源读写支持，如 Spark SQL 提供了很方便的数据获取机制，可以从 Hive 表、外部数据库（JDBC）、RDD、Parquet 文件、JSON 文件等获取数据。第二，Spark SQL 提供了直接访问列的能力。

因此，Spark SQL 是一个既支持 SQL 式又支持命令式数据处理的工具，主要适用场景是处理结构化数据和较为规范的半结构化数据。

Spark SQL 是 Spark 用来处理结构化数据的一个模块。Spark SQL 还提供了多种使用方式，包括 DataFrame API 和 Dataset API。但无论哪种 API 或者编程语言，它们都基于同样的执行引擎，因此我们可以在不同的 API 之间随意切换，它们各有各的特点，在实际开发应用中我们根据业务场景来选择适合的方式即可。

8.4 Spark SQL 特点

Spark SQL 主要有以下 4 个特点。

- 易整合：可以使用 Java、Scala、Python、R 语言的 API 操作。
- 统一的数据访问：连接到任何数据源的方式相同。
- 兼容 Hive：支持 HiveQL 的语法，兼容 Hive（元数据库、SQL 语法、UDF、序列化、反序列化机制）。
- 标准的数据连接：可以使用行业标准的 JDBC 或 ODBC 连接。

8.5 Spark SQL 数据抽象

Spark SQL 增加了 DataFrame（即带有 Schema 信息的 RDD），使开发者可以在 Spark SQL 中执行 SQL 语句。数据既可以来自 RDD，也可以来自 Hive、HDFS、Cassandra 等外部数据源，还可以是 JSON 格式的数据。

Spark SQL 目前支持 Scala、Java、Python 语言，以及支持 SQL-92 规范，如图 8-2 所示。

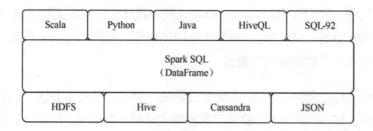

图 8-2

8.6 DataFrame 简介

DataFrame 的前身是 SchemaRDD，从 Spark 1.3.0 开始，SchemaRDD 更名为 DataFrame，并不再直接继承 RDD，而是自己实现 RDD 的绝大多数功能。DataFrame 是一种以 RDD 为基础的分布式数据集，类似于传统数据库的二维表，带有 Schema 元信息（可以理解为数据库的列名和类型）。

DataFrame 可以看作分布式 Row 对象的集合，提供了由列组成的详细 Schema 模式信息，相比 RDD 提供了更多的算子，还可以对 DataFrame 进行执行计划的优化操作。

Spark 能够以二进制的形式序列化数据（不包括数据结构 Schema 信息）到堆外内存（off-heap）中，当要操作数据时，直接操作 off-heap。off-heap 类似于地盘，Schema 类似于地图，Spark 有地图又有自己的地盘，就可以自己说了算，不再受 JVM 的限制，也就不再受 JVM 垃圾回收机制的困扰。通过 Schema 和 off-heap，DataFrame 弥补了 RDD 的缺陷。DataFrame 拥有属于自己的新的执行引擎 Tungsten 和新的语法解析框架 Catalyst。

DataFrame 弥补了 RDD 的缺陷，但是丢了 RDD 的优点。DataFrame 不是类型安全的，其 API 也不是面向对象的。

```
// API不是面向对象的
idAgeDF.filter(idAgeDF.col("age") > 25)
// 不会报错，DataFrame不是类型安全的
idAgeDF.filter(idAgeDF.col("age") > "")
```

这迫使 DataFrame 继续向前发展。

8.7 Dataset 简介

Dataset 是在 Spark 1.6 中引入的 API。与 RDD 相比，Dataset 保存了更多的描述信息，概念上等同于关系数据库中的二维表。与 DataFrame 相比，Dataset 保存了类型信息，是强类型的，提供了编译时类型检查，调用 Dataset 的方法会先生成逻辑计划，然后被 Spark 优化器优化，最终生成物理计划，提交到集群中运行。

Dataset 包含 DataFrame 的功能，在 Spark 2.0 中两者统一。DataFrame 表示为

Dataset[Row]，即 Dataset 的子集，如图 8-3 所示。

```
package object sql {

  /**
   * Converts a logical plan into zero or
   * with the query planner and is not de
   * writing libraries should instead con
   * [[org.apache.spark.sql.sources]]
   */
  @DeveloperApi
  @InterfaceStability.Unstable
  type Strategy = SparkStrategy

  type DataFrame = Dataset[Row]
}
```

图 8-3

Dataset 不同于 RDD，没有使用 Java 序列化器或者 Kryo 进行序列化，而是使用一个特定的编码器进行序列化，这些序列化器可以自动生成，而且在 Spark 执行很多操作（过滤、排序等）的时候不用进行反序列化。

- 编译时的类型安全检查：性能大幅提升，内存使用率大幅降低；极大减少网络数据传输量，极大降低 Scala 和 Java 的代码差异性。
- DataFrame 每一行对应一列。而 Dataset 的定义更加宽松，每一条记录对应一个任意的类型。DataFrame 只是 Dataset 的一种特例。
- Dataset 以 Catalyst 逻辑执行计划，并且数据以编码的二进制形式被存储，不需要反序列化就可以执行过滤、排序等操作。
- Dataset 创建需要一个显式的编码器，把对象序列化为二进制。

Dataset 是特殊的 DataFrame，而 DataFrame 是特殊的 RDD，Dataset 是一个分布式的表。

8.8 RDD、DataFrame 和 Dataset 的区别

Spark 第一代 API 是 RDD，第二代 API 是 DataFrame，第三代 API 是 Dataset，其演化过程如图 8-4 所示。

RDD[Person] 以 Person 为类型参数，但不了解其内部结构。DataFrame 提供了详细的结构信息（Schema 列的名称和类型），这样看起来就像一张表了。Dataset 不光有 Schema 信息，还有类型信息。

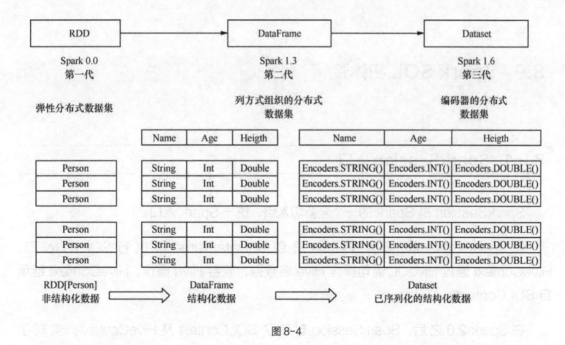

图 8-4

假设 RDD 中的两行数据是这样的——RDD[Person]，如图 8-5 所示。

那么 DataFrame 中的数据就是这样的——DataFrame = RDD[Person] - 泛型 + Schema + SQL 操作 + 优化，如图 8-6 所示。

| 1, 张三, 23 |
| 2, 李四, 35 |

图 8-5

ID:String	Name:String	Age:Int
1	张三	23
2	李四	35

图 8-6

那么 Dataset 中的数据就是这样的——Dataset[Person] == DataFrame + 泛型，如图 8-7 所示。

Dataset 中的数据也可能是这样的——Dataset[Row]

DataFrame == Dataset[Row]，如图 8-8 所示。

value:People[age: bigint, id: bigint, name:string]
People(id=1, name="张三", age=23)
People(id=1, name="李四", age=35)

图 8-7

value:String
1, 张三, 23
2, 李四, 35

图 8-8

8.9 Spark SQL 初体验

8.9.1 SparkSession 入口

SparkSession 是 Spark 的一个全新切入点，统一 Spark 入口。

在 Spark 2.0 之前，SQLContext 是创建 DataFrame 和执行 SQL 的入口。HiveContext 通过 HiveQL 语句操作 Hive 表数据，兼容 Hive 操作，HiveContext 继承自 SQLContext。

在 Spark 2.0 之后，SparkSession 封装了 SQLContext 及 HiveContext，实现了 SQLContext 及 HiveContext 的所有功能。通过 SparkSession 还可以获取 SparkContext。

在 spark-shell 中 SparkSession 和 SparkContext 都已创建，如图 8-9 所示。

```
[hadoop@master bin]$ ./spark-shell --master spark://master:7077 --executor-memory 1g --total-executor-cores 2
Setting default log level to "WARN".
To adjust logging level use sc.setLogLevel(newLevel). For SparkR, use setLogLevel(newLevel).
21/09/10 04:52:04 WARN util.NativeCodeLoader: Unable to load native-hadoop library for your platform... using bu
iltin-java classes where applicable
21/09/10 04:52:33 WARN metastore.ObjectStore: Failed to get database global_temp, returning NoSuchObjectExceptio
n
Spark context Web UI available at http://192.168.52.102:4040
Spark context available as 'sc' (master = spark://master:7077, app id = app-20210910045207-0000).
Spark session available as 'spark'.
Welcome to
      ____              __
     / __/__  ___ _____/ /__
    _\ \/ _ \/ _ `/ __/  '_/
   /___/ .__/\_,_/_/ /_/\_\   version 2.2.0
      /_/

Using Scala version 2.11.8 (Java HotSpot(TM) 64-Bit Server VM, Java 1.8.0_144)
Type in expressions to have them evaluated.
Type :help for more information.
```

图 8-9

在程序当中这样创建 SparkSession 对象，如下所示。

```
//创建SparkSession对象
val spark = SparkSession
  .builder
  .appName("test")
  .enableHiveSupport()
  .getOrCreate()
```

```
// 创建之后可设置运行参数
spark.conf.set("spark.sql.shuffle.partitions", 6)
spark.conf.set("spark.executor.memory", "2g")
```

8.9.2 创建 DataFrame

创建 DataFrame 相当于创建带有 Schema 信息的 RDD，Spark 通过 Schema 信息就能够读懂数据。DataFrame 提供了详细的数据结构信息，从而使得 Spark SQL 可以清楚地知道该数据集中包含哪些列、每列的名称和类型各是什么、DataFrame 中的数据结构信息（即 Schema 信息）。

1. 读取文本文件

① 在本地创建一个文件，有 id、name、age 这 3 列，用空格分隔。

```
vim /home/hadoop/person.txt
zhangsan 20
lisi 29
wangwu 25
zhaoliu 30
sunqi 35
zhouba 40
```

② 上传数据文件到 HDFS。

```
hadoop fs -put /home/hadoop/person.txt /
```

③ 在 spark-shell 中执行下面的命令读取数据，将每一行的数据使用列分隔符分隔。

```
/export/servers/spark/bin/spark-shell \
--master spark://master:7077 \
--executor-memory 1g \
--total-executor-cores 2
```

④ 创建 RDD。

```
val lineRDD= sc.textFile("hdfs://master:9000/person.txt").map(_.split(" "))
//输出结果类型：RDD[Array[String]]
```

⑤ 定义 case class（相当于表的 Schema）。

```
case class Person(id:Int, name:String, age:Int)
```

⑥ 将 RDD 和 case class 关联。

```
val personRDD = lineRDD.map(x => Person(x(0).toInt, x(1), x(2).toInt))
//输出结果类型：RDD[Person]
```

⑦ 将 RDD 转换成 DataFrame。

```
val personDF = personRDD.toDF
//输出结果类型：DataFrame
```

⑧ 查看数据和 Schema 信息。

```
personDF.show
```

结果如图 8-10 所示。

⑨ 查看 personDF 的模式信息，相当于数据库表的元信息。

```
personDF.printSchema
```

结果如图 8-11 所示。

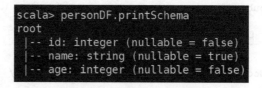

图 8-10　　　　　　　　　　　　　图 8-11

2. 注册表

将数据注册为表数据。

```
personDF.createOrReplaceTempView("t_person")
```

3. 执行 SQL

在表上执行 SQL 查询语句。

```
spark.sql("select id,name from t_person where id > 3").show
```

结果如图 8-12 所示。

```
scala> spark.sql("select id, name from t_person where id>3").show
+---+-------+
| id|   name|
+---+-------+
|  4|zhaoliu|
|  5| tianqi|
|  6|   kobe|
+---+-------+
```

图 8-12

也可以通过 SparkSession 构建 DataFrame。

```
val dataFrame=spark.read.text("hdfs://master:9000/person.txt")
dataFrame.show
```

结果如图 8-13 所示。

4. 读取 JSON 文件

① 使用 Spark 安装包下的 JSON 文件。

```
more /home/hadoop/spark/examples/src/main/resources/people.json
{"name":"Michael"}
{"name":"Andy", "age":30}
{"name":"Justin", "age":19}
```

② 在 spark-shell 中执行下面的命令读取数据。

```
val jsonDF= spark.read.json("file:///home/hadoop/spark/examples/src/main/resources/people.json")
```

③ 使用 DataFrame 的函数操作。

```
jsonDF.show
```

结果如图 8-14 所示。

```
scala> dataFrame.show
+-------------+
|        value|
+-------------+
|1 zhangsan 20|
|     2 lisi 29|
|   3 wangwu 25|
|   4 zhaoliu 30|
|    5 tianqi 35|
|      6 kobe 40|
+-------------+
```

图 8-13

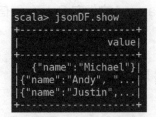

图 8-14

直接读取 JSON 文件有 Schema 信息，因为 JSON 文件本身含有 Schema 信息，Spark SQL 可以自动解析这些信息。

5．读取 Parquet 文件

Parquet 是一种流行的列式存储格式，可以高效地存储具有嵌套字段的记录，还可以针对相同类型的列进行压缩。

Parquet 是语言无关的，而且不与任何一种数据处理框架绑定在一起，适配多种语言和组件。能够与 Parquet 配合的组件有以下 3 类。

- 查询引擎：Hive、Impala、Pig、Presto、Drill、Tajo、HAWQ、IBM Db2 Big SQL。
- 计算框架：MapReduce、Spark、Cascading、Crunch、Scalding、Kite。
- 数据模型：Avro、Thrift、Protocol Buffers、POJO。

使用 Spark 安装包下的 Parquet 文件，查看命令如图 8-15 所示。

```
more /home/hadoop /spark/examples/src/main/resources/users.parquet
```

在 spark-shell 中执行下面的命令读取数据。

```
scala>val parquetDF=spark.read.parquet("file:///export/servers/spark/examples/src/main/resources/users.parquet")
scala>parquetDF: org.apache.spark.sql.DataFrame = [name: string, favorite_color: string ... 1 more field]
```

接下来就可以使用 DataFrame 的函数操作，结果如图 8-16 所示。

```
parquetDF.show
```

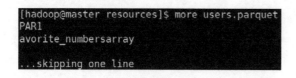
图 8-15

图 8-16

直接读取 Parquet 文件有 Schema 信息，因为 Parquet 文件中保存了列的信息。

8.9.3 创建 Dataset

创建 Dataset 有多种方法，下面主要来学习 3 种创建 Dataset 的方法。

通过 spark.createDataset 方法创建 Dataset，代码如下。

```
//加载数据，读取HDFS中的person.txt文件
val fileRdd = sc.textFile("hdfs://master:9000/person.txt")    //RDD[String]
//调用createDataset方法创建Dataset对象
val ds1 = spark.createDataset(fileRdd)
//DataSet[String] ds1: org.apache.spark.sql.Dataset[String] = [value: string]
ds1.show
```

结果如图 8-17 所示。

通过 toDS 方法生成 Dataset，代码如下。

```
//定义样例类Person
case class Person(name:String, age:Int)
//预备数据
val data = List(Person("zhangsan",20),Person("lisi",30))    //List[Person]
//获取RDD对象
val dataRDD = sc.makeRDD(data)
//调用toDS方法将其转化为Dataset对象
val ds2 = dataRDD.toDS
//ds2: org.apache.spark.sql.Dataset[Person] = [name: string, age: int]
ds2.show
```

结果如图 8-18 所示。

图 8-17

图 8-18

通过 DataFrame.as[泛型] 生成 Dataset，代码如下。

```
//创建Person样例类
case class Person(name:String, age:Int)
//加载people.json文件
val jsonDF= spark.read.json("file:///export/servers/spark/examples/src/main/resources/people.json")    //jsonDF: org.apache.spark.sql.DataFrame = [age: bigint, name: string]
//调用as方法将DataFrame转化为Dataset
val jsonDS = jsonDF.as[Person]
//jsonDS: org.apache.spark.sql.Dataset[People] = [age: bigint, name: string]
```

```
//调用Dataset所提供的show函数查看内容
jsonDS.show
```

结果如图 8-19 所示。

Dataset 也可以注册成表进行查询操作，代码如下。

```
//将Dataset注册为表t_person
jsonDS.createOrReplaceTempView("t_person")
//使用SQL语句进行表查询
spark.sql("select * from t_person").show
```

结果如图 8-20 所示。

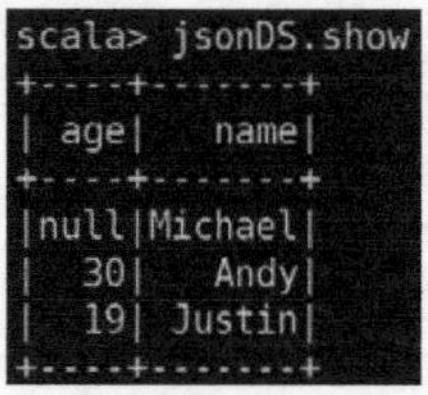

图 8-19

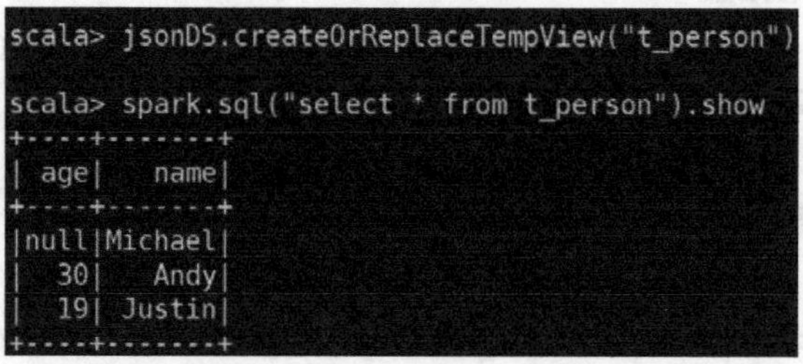

图 8-20

8.9.4　两种查询风格

首先做一些准备工作，读取文件并将其转换为 DataFrame 或 Dataset 数据抽象模型，代码如下。

```
//从HDFS加载数据，并通过map函数映射输出一组数组元素
scala> val lineRDD= sc.textFile("hdfs://master:9000/person.txt").map(_.split(" "))
//创建Person样例类
scala> case class Person(id:Int, name:String, age:Int)
//将RDD转化为Person类型的RDD
scala> val personRDD = lineRDD.map(x => Person(x(0).toInt, x(1), x(2).toInt))
//将RDD转化为DataFrame
scala> val personDF = personRDD.toDF
personDF: org.apache.spark.sql.DataFrame = [id: int, name: string ... 1 more field]
//查看personDF中的内容
personDF.show
//将RDD转化为Dataset
//scala> val personDS = personRDD.toDS
```

```
personDS: org.apache.spark.sql.Dataset[Person] = [id: int, name: string ...
1 more field]
personDS.show
```

结果如图 8-21 和图 8-22 所示。

图 8-21　　　　　　　　图 8-22

1. DSL 风格

Spark SQL 提供了领域特定语言（Domain-Specific Language，DSL）以方便操作结构化数据。通过下面的命令来认识 DSL 风格的查询语言。

示例：查看 name 字段的数据。

```
personDF.select(personDF.col("name")).show
personDF.select(personDF("name")).show
personDF.select(col("name")).show
personDF.select("name").show
```

结果如图 8-23 所示。

示例：查看 name 和 age 字段的数据。

```
personDF.select("name", "age").show
```

结果如图 8-24 所示。

图 8-23　　　　　　　　图 8-24

示例：查询所有 name 和 age 字段的数据，并将 age 字段的数据加 1。

```
personDF.select(personDF.col("name"), personDF.col("age") + 1).show
personDF.select(personDF("name"), personDF("age") + 1).show
personDF.select(col("name"), col("age") + 1).show
personDF.select("name","age").show
//personDF.select("name", "age"+1).show
personDF.select($"name",$"age",$"age"+1).show
```

结果如图 8-25 所示。

示例：过滤 age 大于等于 25 的数据，使用 filter 方法过滤。

```
personDF.filter(col("age") >= 25).show
personDF.filter($"age" >25).show
```

结果如图 8-26 所示。

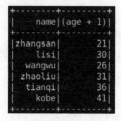

```
+--------+---------+
|    name|(age + 1)|
+--------+---------+
|zhangsan|       21|
|    lisi|       30|
|  wangwu|       26|
| zhaoliu|       31|
|  tianqi|       36|
|    kobe|       41|
+--------+---------+
```

图 8-25

```
+---+-------+---+
| id|   name|age|
+---+-------+---+
|  2|   lisi| 29|
|  3| wangwu| 25|
|  4|zhaoliu| 30|
|  5| tianqi| 35|
|  6|   kobe| 40|
+---+-------+---+
```

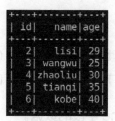

图 8-26

示例：按 age 进行分组并统计相同 age 的人数。

```
personDF.groupBy("age").count().show
```

结果如图 8-27 所示。

示例：统计 age 大于 30 的人数。

```
personDF.filter(col("age")>30).count()
personDF.filter($"age" >30).count()
```

结果如图 8-28 所示。

```
scala> personDF.groupBy("age").count().show
+---+-----+
|age|count|
+---+-----+
| 20|    1|
| 40|    1|
| 35|    1|
| 25|    1|
| 29|    1|
| 30|    1|
+---+-----+
```

图 8-27

```
scala> personDF.filter(col("age")>30).count()
res18: Long = 2
```

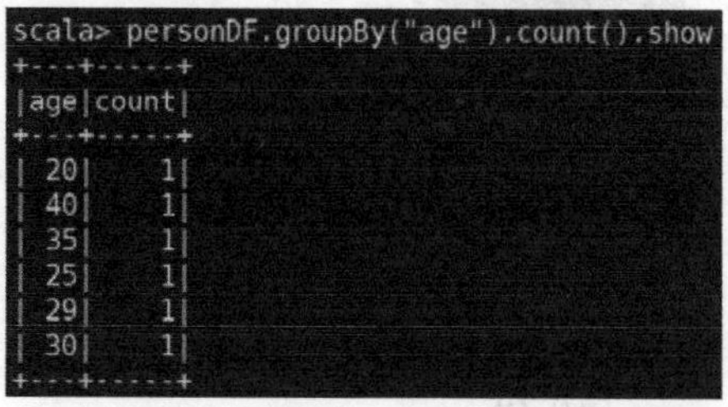

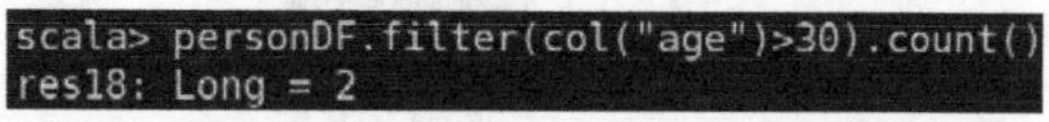

图 8-28

2. SQL 风格

DataFrame 的一个强大之处就是我们可以将它视为一个关系数据表，然后在程序中使用 spark.sql 来执行 SQL 查询，结果将作为一个 DataFrame 返回。如果想使用 SQL 风格的语法，需要将 DataFrame 注册成表。

```
personDF.createOrReplaceTempView("t_person")
spark.sql("select * from t_person").show
```

结果如图 8-29 所示。

示例：显示表的描述信息。

```
spark.sql("desc t_person").show
```

结果如图 8-30 所示。

```
scala> spark.sql("select * from t_person").show
+---+--------+---+
| id|    name|age|
+---+--------+---+
|  1|zhangsan| 20|
|  2|    lisi| 29|
|  3|  wangwu| 25|
|  4| zhaoliu| 30|
|  5|  tianqi| 35|
|  6|    kobe| 40|
+---+--------+---+
```

图 8-29

```
scala> spark.sql("desc t_person").show
+--------+---------+-------+
|col_name|data_type|comment|
+--------+---------+-------+
|      id|      int|   null|
|    name|   string|   null|
|     age|      int|   null|
+--------+---------+-------+
```

图 8-30

示例：查询 age 最大的前两名。

```
spark.sql("select * from t_person order by age desc limit 2").show
```

结果如图 8-31 所示。

```
scala> spark.sql("select * from t_person order by age desc limit 2").show
+---+------+---+
| id|  name|age|
+---+------+---+
|  6|  kobe| 40|
|  5|tianqi| 35|
+---+------+---+
```

图 8-31

示例：查询 age 大于 30 的人的信息。

```
spark.sql("select * from t_person where age > 30 ").show
```

结果如图 8-32 所示。

```
scala> spark.sql("select * from t_person where age > 30 ").show
+---+-----+---+
| id| name|age|
+---+-----+---+
|  5|tianqi| 35|
|  6| kobe| 40|
+---+-----+---+
```

图 8-32

示例：使用 SQL 风格完成 DSL 中的需求。

//查询 name 和 age+1 的信息

spark.sql("select name, age + 1 from t_person").show

结果如图 8-33 所示。

```
scala> spark.sql("select name, age + 1 from t_person").show
+--------+---------+
|    name|(age + 1)|
+--------+---------+
|zhangsan|       21|
|    lisi|       30|
|  wangwu|       26|
| zhaoliu|       31|
|  tianqi|       36|
|    kobe|       41|
+--------+---------+
```

图 8-33

//查询 age 大于 25 的人的信息

spark.sql("select name, age from t_person where age > 25").show

结果如图 8-34 所示。

```
scala> spark.sql("select name, age from t_person where age > 25").show
+-------+---+
|   name|age|
+-------+---+
|   lisi| 29|
|zhaoliu| 30|
| tianqi| 35|
|   kobe| 40|
+-------+---+
```

图 8-34

DataFrame 和 Dataset 可以通过 RDD 创建，也可以通过读取普通文本创建。需要注意的是，直接读取没有完整的约束，需要通过 RDD+Schema。例如我们创建样例类来封装 Schema 信息。通过 JSON 和 Parquet 方式会有完整的约束。不管是 DataFrame 还是 Dataset 都可以注册成表，之后就可以使用 SQL 进行查询，也可以使用 DSL。

8.10 本章总结

本章主要介绍了 Spark SQL 编程的基本概念，其中包含 DataFrame 和 Dataset 数据抽象模型，并介绍了 Spark SQL 的开发案例，读者此时应该初步掌握了 Spark SQL 编程框架的基本应用。

8.11 本章习题

1. 总结 Spark SQL 数据抽象模型 DataFrame 和 Dataset 的特征。
2. 实践 Spark SQL 的编程案例。

第 9 章 Spark SQL 结构化数据处理高级应用

前面我们在 spark-shell 中快速体验了 Spark SQL 在对结构化数据操作时的便捷，接下来我们将在 IDEA 中进行 Spark SQL 程序的开发。

9.1 使用 IDEA 开发 Spark SQL

首先，我们要打开 IDEA（见图 9-1），在其中运行代码之前，先进行本地测试运行，如果没有问题，就可以将代码打包并提交到集群上运行。下面重点讲解如何在 IDEA 中编写 Spark SQL 代码。

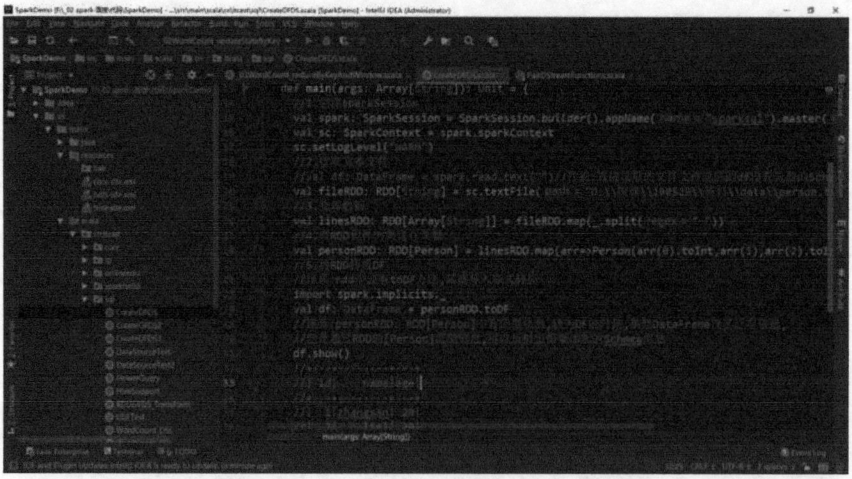

图 9-1

在开发之前需要明确一点，Spark 会根据文件信息试着去推断 DataFrame/Dataset 的 Schema 信息，当然我们也可以手动指定，手动指定的方式有以下 3 种。

- 为指定列名添加 Schema 信息。
- 指定 Schema 信息。
- 编写样例类，利用反射机制推断 Schema 信息。

9.1.1 创建 DataFrame 和 Dataset

下面我们通过在 IDEA 中开发 Spark SQL 来实现为 DataFrame 或者 Dataset 手动指定 Schema 信息的功能。

1. 为指定列名添加 Schema 信息

为指定列名添加 Schema 信息，代码如下。

```
package cn.stonesoup.sql
import org.apache.spark.SparkContext
import org.apache.spark.rdd.RDD
import org.apache.spark.sql.{DataFrame, SparkSession}
/*
         创建DataFrame和Dataset的时候为指定的列名添加Schema信息
*/
object CreateDFDS {
    //main方法，程序的入口
  def main(args: Array[String]): Unit = {
    //1.创建SparkSession
val spark: SparkSession = SparkSession.builder()
.master("local[*]").appName("Spark SQL").getOrCreate()
//获取SparkContext对象
    val sc: SparkContext = spark.sparkContext
    sc.setLogLevel("WARN")
    //2.读取文件
val fileRDD: RDD[String] = sc.textFile("D:\\data\\person.txt")
//处理数据
    val linesRDD: RDD[Array[String]] = fileRDD.map(_.split(" "))
    val rowRDD: RDD[(Int, String, Int)] = linesRDD.map(line =>(line(0).toInt,
line(1),line(2).toInt))
    //3.将RDD转换成DataFrame，并加上Schema信息
```

```
    /*注意：RDD中原本没有toDF方法，新版本中要给它增加一个方法，可以使用隐式转换
import spark.implicits._*/
    val personDF: DataFrame = rowRDD.toDF("id","name","age")
    personDF.show(10)  //最多显示10行数据
    personDF.printSchema()
    sc.stop()
    spark.stop()
  }
}
```

运行结果如下所示。

```
+---+--------+---+
| id|    name|age|
+---+--------+---+
|  1|zhangsan| 20|
|  2|    lisi| 29|
|  3|  wangwu| 25|
|  4| zhaoliu| 30|
|  5|  tianqi| 35|
|  6|    kobe| 40|
+---+--------+---+
```

2. 利用反射机制推断 Schema 信息

利用反射机制推断 Schema 信息，代码如下。

```
package cn.stonesoup.sql
import org.apache.spark.SparkContext
import org.apache.spark.rdd.RDD
import org.apache.spark.sql.{DataFrame, SparkSession}
object CreateDFDS3 {
  def main(args: Array[String]): Unit = {
    //1.创建SparkSession
val spark: SparkSession = SparkSession.builder()
.master("local[*]").appName("Spark SQL").getOrCreate()
//获取SparkContext 对象
    val sc: SparkContext = spark.sparkContext
    sc.setLogLevel("WARN")
    //2.读取文件
val fileRDD: RDD[String] = sc.textFile("D:\\data\\person.txt")
```

```
//处理数据
val linesRDD: RDD[Array[String]] = fileRDD.map(_.split(" "))
//将RDD和样例类进行关联
val rowRDD: RDD[Person] = linesRDD
.map(line =>Person(line(0).toInt,line(1),line(2).toInt))
    //3.将RDD转换成DataFrame
    /*注意：RDD中原本没有toDF方法，新版本中要给它增加一个方法，可以使用隐式转换
    import spark.implicits._*/
    //注意：上面的rowRDD的泛型是Person，里面包含Schema信息
    //所以Spark SQL可以通过反射自动获取Schema信息并将其添加给DataFrame
    val personDF: DataFrame = rowRDD.toDF
    personDF.show(10)
    personDF.printSchema()
    sc.stop()
    spark.stop()
  }
  case class Person(id:Int,name:String,age:Int)
}
```

运行结果如下所示。

```
+---+--------+---+
| id|    name|age|
+---+--------+---+
|  1|zhangsan| 20|
|  2|    lisi| 29|
|  3|  wangwu| 25|
|  4| zhaoliu| 30|
|  5|  tianqi| 35|
|  6|    kobe| 40|
+---+--------+---+
```

9.1.2 花式查询

我们可以在 IDEA 中编程，实现 Spark SQL 的 SQL 和 DSL 花式查询，如查询所有数据，查询比 age 长一岁的人，查询 age 最大的两个人，查询各个年龄阶段的人数，查询 age 大于 30 的人数等，代码如下。

```
package cn.stonesoup.sql
import org.apache.spark.SparkContext
```

```scala
import org.apache.spark.rdd.RDD
import org.apache.spark.sql.{DataFrame, SparkSession}


object QueryDemo {
  def main(args: Array[String]): Unit = {
    //1.创建SparkSession
    val spark: SparkSession = SparkSession
.builder().master("local[*]").appName("Spark SQL").getOrCreate()
    val sc: SparkContext = spark.sparkContext
    sc.setLogLevel("WARN")
    //2.读取文件
    val fileRDD: RDD[String] = sc.textFile("D:\\data\\person.txt")
     //处理数据
    val linesRDD: RDD[Array[String]] = fileRDD.map(_.split(" "))
    //RDD关联样例类Person
    val rowRDD: RDD[Person] = linesRDD.map(line =>Person(line(0).toInt,line(1),line(2).toInt))
    //3.将RDD转换成DataFrame
    /*注意：RDD中原本没有toDF方法，新版本中要给它增加一个方法，可以使用隐式转换
    import spark.implicits._ */
    //注意：上面的rowRDD的泛型是Person，里面包含Schema信息
    //所以Spark SQL可以通过反射自动获取Schema信息并将其添加给DataFrame
    val personDF: DataFrame = rowRDD.toDF
    personDF.show(10)
    personDF.printSchema()
    //========================SQL方式查询========
    //注册表
    personDF.createOrReplaceTempView("t_person")
    //1.查询所有数据
    spark.sql("select * from t_person").show()
    //2.查询比age长一岁的人
    spark.sql("select age,age+1 from t_person").show()
    //3.查询age最大的两个人
    spark.sql("select name,age from t_person order by age desc limit 2").show()
    //4.查询各个年龄阶段的人数
    spark.sql("select age,count(*) from t_person group by age").show()
    //5.查询age大于30的人数
    spark.sql("select * from t_person where age > 30").show()
```

```
//========================DSL方式查询========
//1.查询所有数据
personDF.select("name","age")
//2.查询比age长一岁的人
personDF.select($"name",$"age" + 1)
//3.查询age最大的两个人
personDF.sort($"age".desc).show(2)
//4.查询各个年龄阶段的人数
personDF.groupBy("age").count().show()
//5.查询age大于30的人数
personDF.filter($"age" > 30).show()

    sc.stop()
    spark.stop()
  }
  case class Person(id:Int,name:String,age:Int)
}
```

9.1.3 相互转换

Spark SQL 开发特别灵活,使用 RDD、DataFrame 还是 Dataset 要根据业务需求和个人对于 API 的熟练程度进行选择。官方推荐使用 Dataset,RDD 后续可能不再更新,但是 RDD 的算子非常灵活,且在机器学习算法中使用较多。

RDD、DataFrame、Dataset 之间的转换也有很多种(6种),但是我们实际操作就只有两种,如图 9-2 所示。

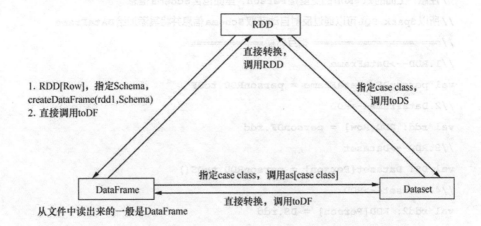

图 9-2

- 使用 RDD 算子操作。
- 使用 DSL/SQL 对表操作。

```scala
package cn.stonesoup.sql
import org.apache.spark.SparkContext
import org.apache.spark.rdd.RDD
import org.apache.spark.sql.{DataFrame, Dataset, Row, SparkSession}
/*
            实现RDD、DataFrame以及Dataset三者之间的转换操作
*/
object TransformDemo {
  def main(args: Array[String]): Unit = {
    //1.创建SparkSession
    val spark: SparkSession = SparkSession.builder()
.master("local[*]").appName("Spark SQL").getOrCreate()
    val sc: SparkContext = spark.sparkContext
    sc.setLogLevel("WARN")
    //2.读取文件
    val fileRDD: RDD[String] = sc.textFile("D:\\data\\person.txt")
     //处理数据
    val linesRDD: RDD[Array[String]] = fileRDD.map(_.split(" "))
    //RDD关联样例类Person
    val personRDD: RDD[Person] = linesRDD.map(line =>Person(line(0).toInt,line(1),line(2).toInt))
    //3.将RDD转换成DataFrame
    //注意：RDD中原本没有toDF方法，新版本中要给它增加一个方法,可以使用隐式转换
    import spark.implicits._
    //注意：上面的rowRDD的泛型是Person，里面包含Schema信息
    // 所以Spark SQL可以通过反射自动获取Schema信息并将其添加给DataFrame
    //====================相互转换===========
    //1.RDD-->DataFrame
    val personDF: DataFrame = personRDD.toDF
    //2.DataFrame-->RDD
    val rdd: RDD[Row] = personDF.rdd
    //3.RDD-->Dataset
    val DS: Dataset[Person] = personRDD.toDS()
    //4.Dataset-->RDD
    val rdd2: RDD[Person] = DS.rdd
    //5.DataFrame-->Dataset
```

```
    val DS2: Dataset[Person] = personDF.as[Person]
    //6.Dataset-->DataFrame
    val DF: DataFrame = DS2.toDF()

    sc.stop()
    spark.stop()
  }
  case class Person(id:Int,name:String,age:Int)
}
```

9.1.4 Spark SQL 词频统计实战

词频统计是学习大数据的一个经典案例，下面通过两种 Spark SQL 风格来实现其功能。

1. SQL 风格

```
package cn.stonesoup.sql
import org.apache.spark.SparkContext
import org.apache.spark.sql.{DataFrame, Dataset, SparkSession}

object WordCount {
  def main(args: Array[String]): Unit = {
    //1.创建SparkSession
    val spark: SparkSession = SparkSession.builder()
.master("local[*]").appName("Spark SQL").getOrCreate()
    //获取SparkContext对象
    val sc: SparkContext = spark.sparkContext
    sc.setLogLevel("WARN")
    //2.读取文件
    val fileDF: DataFrame = spark.read.text("D:\\data\\words.txt")
    val fileDS: Dataset[String] = spark.read.textFile("D:\\data\\words.txt")
    //fileDF.show()
    //fileDS.show()
    //3.对每一行按照空格进行切分并压平
    //fileDF.flatMap(_.split(" "))
    //注意：错误，因为DataFrame没有泛型，不知道_的类型是String
```

```scala
import spark.implicits._
val wordDS: Dataset[String] = fileDS.flatMap(_.split(" "))
//注意：正确，因为Dataset有泛型，知道_的类型是String
//wordDS.show()
/*
+-----+
|value|
+-----+
|hello|
|   me|
|hello|
|  you|
  ...
*/
//4.对上面的数据进行词频统计
wordDS.createOrReplaceTempView("t_word")
    //按value字段进行分组，统计value出现次数，并按次数倒序排列
val sql =
  """
    |select value ,count(value) as count
    |from t_word
    |group by value
    |order by count desc
  """.stripMargin
spark.sql(sql).show()
sc.stop()
spark.stop()
  }
}
```

运行结果如下所示。

```
+-----+------+
|value|counts|
+-----+------+
|hello|    12|
|   me|     4|
|  you|     4|
|  her|     4|
```

2. DSL 风格

```scala
package cn.stonesoup.sql
import org.apache.spark.SparkContext
import org.apache.spark.sql.{DataFrame, Dataset, SparkSession}

object WordCount2 {
  def main(args: Array[String]): Unit = {
    //1.创建SparkSession
    val spark: SparkSession = SparkSession.builder()
.master("local[*]").appName("Spark SQL").getOrCreate()
    val sc: SparkContext = spark.sparkContext
    sc.setLogLevel("WARN")
    //2.读取文件
    val fileDF: DataFrame = spark.read.text("D:\\data\\words.txt")
    val fileDS: Dataset[String] = spark.read.textFile("D:\\data\\words.txt")
    //fileDF.show()
    //fileDS.show()
    //3.对每一行按照空格进行切分并压平
    //fileDF.flatMap(_.split(" "))
//注意: 错误,因为DataFrame没有泛型,不知道_的类型是String
    import spark.implicits._
    val wordDS: Dataset[String] = fileDS.flatMap(_.split(" "))
//注意: 正确,因为Dataset有泛型,知道_的类型是String
    //wordDS.show()
    /*
    +-----+
    |value|
    +-----+
    |hello|
    |   me|
    |hello|
    |  you|
     ...
    */
    //4.对上面的数据进行词频统计
    wordDS.groupBy("value").count().orderBy($"count".desc).show()
    sc.stop()
```

```
    spark.stop()
  }
}
```

运行结果如下所示。

```
+-----+------+
|value|counts|
+-----+------+
|hello|    12|
|   me|     4|
|  you|     4|
|  her|     4|
```

9.2 Spark SQL 多数据源交互

Spark SQL 可以与多种数据源交互，例如普通文本、JSON 文件、Parquet 文件、CSV 文件、MySQL 文件等。

示例：写数据，代码如下。

```
package cn.stonesoup.sql
import java.util.Properties
import org.apache.spark.SparkContext
import org.apache.spark.rdd.RDD
import org.apache.spark.sql.{DataFrame, SaveMode, SparkSession}
    /*
        读取本地文件，并将其写入不同的数据源，如普通文本、JSON文件、Parquet文件等
     */
object WriterDataSourceDemo {
  def main(args: Array[String]): Unit = {
    //1.创建SparkSession
    val spark: SparkSession = SparkSession.builder()
.master("local[*]").appName("Spark SQL").getOrCreate()
    val sc: SparkContext = spark.sparkContext
    sc.setLogLevel("WARN")
    //2.读取文件
    val fileRDD: RDD[String] = sc.textFile("D:\\data\\person.txt")
      //处理数据
```

```scala
    val linesRDD: RDD[Array[String]] = fileRDD.map(_.split(" "))
      //RDD与样例类Person相关联
    val rowRDD: RDD[Person] = linesRDD.map(line =>Person(line(0).toInt,line(1),
line(2).toInt))
    //3.将RDD转换成DataFrame
    /*注意:RDD中原本没有toDF方法,新版本中要给它增加一个方法,可以使用隐式转换
    import spark.implicits._*/
    //注意:上面的rowRDD的泛型是Person,里面包含Schema信息
    //所以Spark SQL可以通过反射自动获取Schema信息并将其添加给DataFrame
    val personDF: DataFrame = rowRDD.toDF
    //==================将DataFrame写入不同数据源============
    //Text data source supports only a single column, and you have 3 columns.;
    //注意:错误,因为写入文本信息时,只支持单列,此时有3列就不行了
    //personDF.write.text("D:\\data\\output\\text")
    //写入JSON文件数据源
    personDF.write.json("D:\\data\\output\\json")
    //写入CSV文件数据源
    personDF.write.csv("D:\\data\\output\\csv")
    //写入Parquet文件数据源
    personDF.write.parquet("D:\\data\\output\\parquet")
    val prop = new Properties()
    prop.setProperty("user","root")
    prop.setProperty("password","root")
      //写入MySQL数据库数据源
    personDF.write.mode(SaveMode.Overwrite).jdbc(
"jdbc:mysql://localhost:3306/bigdata?characterEncoding=UTF-8","person",prop)
    println("写入成功")
    sc.stop()
    spark.stop()
  }
  case class Person(id:Int,name:String,age:Int)
}
```

示例:读数据,代码如下。

```scala
package cn.stonesoup.sql
import java.util.Properties
import org.apache.spark.SparkContext
import org.apache.spark.sql.SparkSession
```

```scala
object ReadDataSourceDemo {
  def main(args: Array[String]): Unit = {
    //1.创建SparkSession
    val spark: SparkSession = SparkSession.builder()
.master("local[*]").appName("Spark SQL").getOrCreate()
    val sc: SparkContext = spark.sparkContext
    sc.setLogLevel("WARN")
    //2.读取文件
    //读取JSON文件，并展示
    spark.read.json("D:\\data\\output\\json").show()
    //读取CSV文件，并展示
    spark.read.csv("D:\\data\\output\\csv").toDF("id","name","age").show()
    //读取Parquet文件，并展示
    spark.read.parquet("D:\\data\\output\\parquet").show()
    //访问MySQL数据库所需要的配置信息
    val prop = new Properties()
    prop.setProperty("user","root")
    prop.setProperty("password","root")
    spark.read.jdbc(
"jdbc:mysql://localhost:3306/bigdata?characterEncoding=UTF-8","person",prop).show()
    sc.stop()
    spark.stop()
  }
```

9.3 Spark SQL 自定义函数

9.3.1 自定义函数分类

类似于 Hive 中的自定义函数，在 Spark SQL 中，如果内置函数不够使用，同样可以使用自定义函数来实现某些功能。Spark SQL 中的自定义函数有3类，常用的是如下两类。

- UDF（User-Defined Function，用户自定义函数）：最基本的自定义函数，类似 to_char、to_date 等，输入一行，输出一行。

- UDAF（User-Defined Aggregate Function，用户自定义聚合函数）：类似在 group by 之后使用的 sum、avg 等，输入多行，输出一行。

9.3.2 UDF

udf.txt 数据内容如下。

```
Hello
abc
study
small
……
```

通过 UDF 可以将每一行数据转换成大写，SQL 语句如下。

```
select value,smallToBig(value) from t_word
```

通过如下代码来实现 UDF。

```
package cn.stonesoup.sql
import org.apache.spark.SparkContext
import org.apache.spark.sql.{Dataset, SparkSession}

object UDFDemo {
  def main(args: Array[String]): Unit = {
    //1.创建SparkSession
    val spark: SparkSession = SparkSession.builder()
.master("local[*]").appName("Spark SQL").getOrCreate()
    val sc: SparkContext = spark.sparkContext
    sc.setLogLevel("WARN")
    //2.读取文件
    val fileDS: Dataset[String] = spark.read.textFile("D:\\data\\udf.txt")
    fileDS.show()
    /*
    +----------+
    |     value|
    +----------+
    |helloworld|
    |       abc|
```

```
   |     study|
   | smallWORD|
   +----------+
 */
/*
将每一行数据转换成大写
select value,smallToBig(value) from t_word
*/
//注册一个函数，名称为smallToBig，功能是传入一个字符串，返回一个大写的字符串
spark.udf.register("smallToBig",(str:String) => str.toUpperCase())
fileDS.createOrReplaceTempView("t_word")
//使用自定义的函数
spark.sql("select value,smallToBig(value) from t_word").show()
/* 代码运行结果如下
+----------+--------------------+
|     value|UDF:smallToBig(value)|
+----------+--------------------+
|helloworld|          HELLOWORLD|
|       abc|                 ABC|
|     study|               STUDY|
| smallWORD|           SMALLWORD|
+----------+--------------------+
 */
    sc.stop()
    spark.stop()
  }
}
```

9.3.3 UDAF

有 udaf.json 数据内容如下，求平均薪资是多少。

```
{"name":"Michael","salary":3000}
{"name":"Andy","salary":4500}
{"name":"Justin","salary":3500}
{"name":"Berta","salary":4000}
```

实现 UDAF 需要继承 UserDefinedAggregateFunction 类并重写类中的方法。

- inputSchema：输入数据的类型。
- bufferSchema：产生中间结果的数据类型。
- dataType：最终返回的结果类型。
- deterministic：确保一致性，一般用 true。
- initialize：指定初始值。
- update：每有一条数据参与运算，就更新一下中间结果（update 相当于在每一个分区中的运算）。
- merge：全局聚合 (将每个分区的结果进行聚合)。
- evaluate：计算最终的结果。

通过如下代码来实现 UDAF，完成求平均薪资的需求。

```
package cn.stonesoupsql
import org.apache.spark.SparkContext
import org.apache.spark.sql.expressions.{MutableAggregationBuffer, UserDefinedAggregateFunction}
import org.apache.spark.sql.types._
import org.apache.spark.sql.{DataFrame, Row, SparkSession}

            /*
            通过UDAF来解决计算平均薪资的问题
            */
object UDAFDemo {
  def main(args: Array[String]): Unit = {
    //1.创建SparkSession
    val spark: SparkSession = SparkSession.builder().appName("Spark SQL").master("local[*]").getOrCreate()
    val sc: SparkContext = spark.sparkContext
    sc.setLogLevel("WARN")
    //2.读取文件
    val employeeDF: DataFrame = spark.read.json("D:\\data\\udaf.json")
    //3.创建临时表
    employeeDF.createOrReplaceTempView("t_employee")
    //4.注册UDAF
    spark.udf.register("myavg",new MyUDAF)
    //5.使用UDAF
    spark.sql("select myavg(salary) from t_employee").show()
    //6.使用内置的avg函数
    spark.sql("select avg(salary) from t_employee").show()
  }
```

```scala
}
class MyUDAF extends UserDefinedAggregateFunction{
  //输入数据类型的Schema
  override def inputSchema: StructType = {
    StructType(StructField("input",LongType)::Nil)
  }
  //缓冲区数据类型的Schema,就是转换之后的数据的Schema
  override def bufferSchema: StructType = {
    StructType(StructField("sum",LongType)::StructField("total",LongType)::Nil)
  }
  //返回值的数据类型
  override def dataType: DataType = {
    DoubleType
  }
  //确定是否相同的输入会有相同的输出
  override def deterministic: Boolean = {
    true
  }
  //初始化内部数据结构
  override def initialize(buffer: MutableAggregationBuffer): Unit = {
    buffer(0) = 0L
    buffer(1) = 0L
  }
  //更新内部数据结构,区内计算
  override def update(buffer: MutableAggregationBuffer, input: Row): Unit = {
    //所有的金额相加
    buffer(0) = buffer.getLong(0) + input.getLong(0)
    //一共有多少条数据
    buffer(1) = buffer.getLong(1) + 1
  }
  //对来自不同分区的数据进行合并,全局合并
  override def merge(buffer1: MutableAggregationBuffer, buffer2: Row): Unit = {
    buffer1(0) =buffer1.getLong(0) + buffer2.getLong(0)
    buffer1(1) = buffer1.getLong(1) + buffer2.getLong(1)
  }
  //计算输出数据值
  override def evaluate(buffer: Row): Any = {
    buffer.getLong(0).toDouble / buffer.getLong(1)
  }
}
```

9.4 Spark on Hive

Hive 的元数据库中描述了有哪些数据库、表，表有多少列，每一列是什么类型，以及表的数据保存在 HDFS 的什么位置。当执行 HQL 时，先到 MySQL 元数据库中查找描述信息，然后解析 HQL 并根据描述信息生成 MapReduce 任务。

简单来说，Hive 就是将 SQL 根据 MySQL 中的元数据信息转换成 MapReduce 执行，但是速度慢。使用 Spark SQL 整合 Hive 其实就是让 Spark SQL 去加载 Hive 的元数据库，然后通过 Spark SQL 执行引擎去操作 Hive 表。所以首先需要开启 Hive 的元数据库服务，让 Spark SQL 能够加载元数据。

在 Spark 2.0 之后，SparkSession 对 HiveContext 和 SQLContext 进行了统一，开发者可以通过操作 SparkSession 来操作 HiveContext 和 SQLContext。

9.4.1 开启 Hive 的元数据库服务

① 修改 hive/conf/hive-site.xml，新增如下配置。

```xml
<?xml version="1.0"?>
<?xml-stylesheet type="text/xsl" href="configuration.xsl"?>
<configuration>
    <property>
        <name>hive.metastore.warehouse.dir</name>
        <value>/user/hive/warehouse</value>
    </property>
    <property>
        <name>hive.metastore.local</name>
        <value>false</value>
    </property>
    <property>
        <name>hive.metastore.uris</name>
        <value>thrift://master:9083</value>
    </property>
</configuration>
```

② 在后台开启 Hive 的元数据库服务。

```
nohup /export/servers/hive/bin/hive --service metastore 2>&1 >> /var/log.log &
```

9.4.2　Spark SQL 整合 Hive 元数据库

Spark 有一个内置的元数据库，使用 Derby 嵌入式数据库保存数据，但是这种方式不适合生产环境，因为这种方式同一时间只能使用一个 SparkSession。生产环境中更推荐使用 Hive 元数据库。

Spark SQL 整合 Hive 元数据库的主要思路就是通过配置能够访问它，并且能够使用 HDFS 保存 Warehouse，所以可以直接复制 Hadoop 和 Hive 的配置文件到 Spark 的配置目录。

- hive-site.xml：元数据库的位置等信息。
- core-site.xml：安全相关的配置。
- hdfs-site.xml：HDFS 相关的配置。

使用 IDEA 进行本地测试的话，直接把以上配置文件放在 resource 目录即可。

9.4.3　使用 Spark SQL 操作 Hive 表

通过 Scala 语言实现 Spark SQL 编程操作 Hive 表。

```
package cn.stonesoup.sql
import org.apache.spark.sql.SparkSession

object HiveSupport {
  def main(args: Array[String]): Unit = {
    //创建SparkSession
    val spark = SparkSession.builder().appName("HiveSupport").master("local[*]")
      //.config("spark.sql.warehouse.dir", "hdfs://master:9000/user/hive/warehouse")
      //.config("hive.metastore.uris", "thrift://master:9083")
      .enableHiveSupport()//开启Hive语法的支持
      .getOrCreate()
    spark.sparkContext.setLogLevel("WARN")
    //查看有哪些表
    spark.sql("show tables").show()
    //创建表
```

```
        spark.sql("CREATE TABLE person (id int, name string, age int) row format
delimited fields terminated by ' '")
        //加载数据，数据为当前SparkDemo目录下的person.txt(和src平级)
        spark.sql("LOAD DATA LOCAL INPATH 'SparkDemo/person.txt' INTO TABLE person")
        //查询数据
        spark.sql("select * from person ").show()
        spark.stop()
    }
}
```

9.5 本章总结

本章重点介绍了如何使用 IDEA 开发 Spark SQL 应用程序、Spark SQL 多数据源交互的方式、Spark SQL 自定义函数的开发以及 Spark on Hive 的开发步骤等。

9.6 本章习题

1. 在 IDEA 工具里开发 Spark SQL 应用程序，完成词频统计。

2. 编写 Spark SQL 自定义函数，实现 count 函数的功能。

3. 练习 Spark on Hive 的开发步骤。

第 10 章 Spark Streaming 核心编程

10.1 场景需求

在京东和淘宝等商城上，当我们把一款手机放入购物车的时候，商城会立刻推荐其他热卖型号的手机。这就是商品实时推荐。不难想象，每当用户购买一个商品时如果都要实现实时推荐，数据量将十分巨大。为了准确、快速地推荐商品，需要使用一些推荐算法。

我们再来看一个熟悉的场景，即北京的实时路况，当开车出行时，很多人一般会先打开地图查看一下路况信息。这背后，一般都是几百上千台的机器在实时采集数据、处理数据、推送数据，因此汇报的数据量很大。

一般的大型集群和平台，除需要对 MySQL、HBase 等数据库进行监控，还要针对 Tomcat、Nginx、Node.js 等应用进行监控。往往一个用户的访问行为会带来几百条日志，这些都要汇报，数据量非常大，因此要从这些日志中聚合系统运行状况，以保证业务正常运行，如图 10-1 和图 10-2 所示。

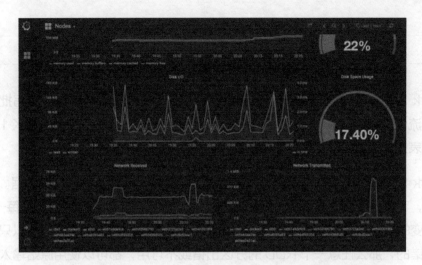

图 10-1

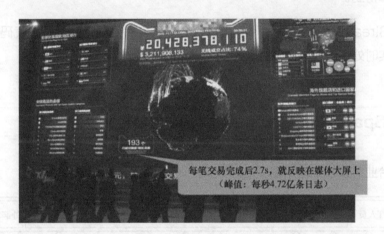

图 10-2

10.2 Spark Streaming 概述

Spark Streaming 是一个基于 SparkCore 的实时计算框架,其使构建可伸缩的容错流应用程序变得容易。其基本原理是:实时输入数据流经 Spark 引擎以类似批处理的方式处理,如图 10-3 所示。

图 10-3

10.2.1　Spark Streaming 的特点

Spark Streaming 将 Spark 的集成 API 引入流处理，让我们可以像编写批处理作业一样编写流作业。它支持 Java、Scala 和 Python 等语言的应用开发。JDK 1.8 后有了 Lambda 表达式，代码读写优雅了很多。

Spark Streaming 在没有额外代码和配置的情况下，可以恢复丢失的数据。对于实时计算来说，容错性至关重要。首先要明确一下 Spark 中 RDD 的容错机制：每一个 RDD 都是不可变的分布式可重算的数据集，它记录着确定性的操作继承关系，所以只要输入数据是可容错的，那么任意一个 RDD 的分区出错或不可用，都可以使用原始输入数据经过转换操作重算得到。

Spark Streaming 可以在 Spark 上运行，并且允许重复使用相同的代码进行批处理。也就是说，实时处理可以与离线处理相结合，实现交互式的查询操作。

10.2.2　Spark Streaming 实时计算所处的位置

数据平台业务整体架构如图 10-4 所示。

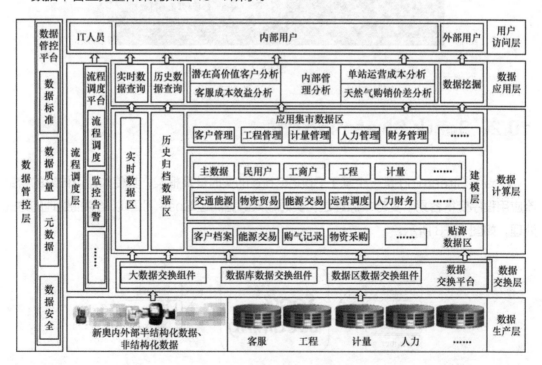

图 10-4

- 数据生产层：生产环境的数据源。
- 数据交换层：负责将生产环境的数据实时或定期更新到数据计算层。
- 数据计算层：负责数据的分类、归档、建模等数据处理操作。
- 数据应用层：负责数据的具体业务应用。
- 用户访问层：负责数据的展示。

数据平台数据处理过程如图 10-5 所示。

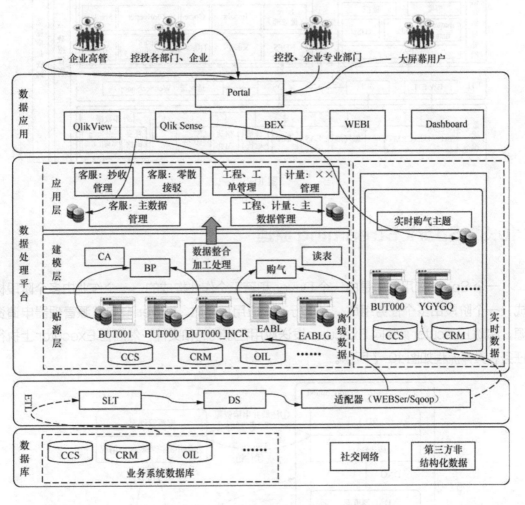

图 10-5

- 贴源层：数据结构、数据与应用服务器系统数据源保持一致，完成数据的增量合并，为数据质量提供检核基础。
- 建模层：依据应用的业务逻辑规则，对缓冲层的数据进行加工、整合，为数据建立连接关系，形成客户、交易等不同的数据主题，为后续的数据使用提供基础。
- 应用层：建立面向各业务领域的分析指标和体系，形成基于各业务分析主题的二维表，如维度、事实表。

数据平台技术架构如图 10-6 所示。

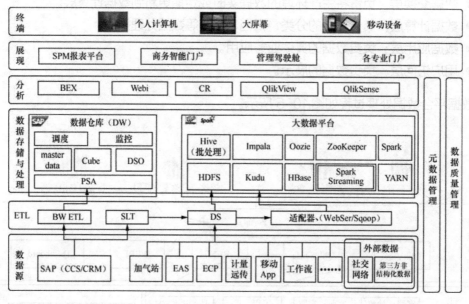

图 10-6

10.3 Spark Streaming 原理

一个 Spark 应用程序是由一个 Driver 和若干个作业构成的。一个作业由多个阶段构成，一个阶段由多个任务组成。当执行一个应用程序时，Driver 会向资源管理器申请资源，启动 Executor，并向 Executor 发送应用程序代码和文件，然后在 Executor 上执行任务，具体流程如图 10-7 所示。

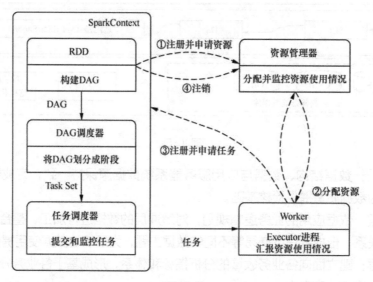

图 10-7

10.3.1 基本流程

在 Spark Streaming 中，会有一个接收器（Receiver）组件，作为一个长期运行的任务运行在一个 Executor 上。Receiver 接收外部的数据流形成 DStream（Discretized Stream，离散化数据流，即连续不断的数据流）。

DStream 会被按照时间间隔划分成一批一批的 RDD，当批处理间隔时间缩短到秒级时，便可以用于处理实时数据流。时间间隔的大小可以由参数指定，一般设在 500ms 到几秒之间。对 DStream 进行操作就是对 RDD 进行操作，计算处理的结果可以传给外部系统。

Spark Streaming 的工作流程如图 10-8 所示。接收到实时数据后，给数据分批次，然后传给 Spark 引擎处理，最后生成该批次的结果。

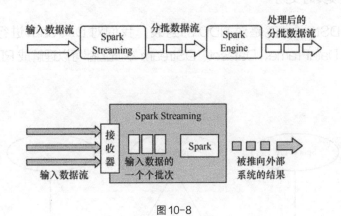

图 10-8

10.3.2 数据模型

Spark Streaming 的基础数据模型是 DStream，代表持续性的数据流和经过各种 Spark 算子操作后的结果数据流。在内部实现上，Spark Streaming 的输入数据按照时间片（如 1s）分成一段一段的，每一段数据转换为 Spark 中的 RDD，这些分段就是 DStream，并且对 DStream 的操作最终都会转变为对相应的 RDD 的操作。

创建 DStream 的方式是：通过文件、Socket、Kafka、Flume 等取得的数据作为输入数据流，或用于其他 DStream 高层操作。在内部，DStream 被表达为 RDD 的一个序列。我们可以从如下多个角度深入理解 DStream。

- DStream 本质上就是一系列时间上连续的 RDD，如图 10-9 所示。

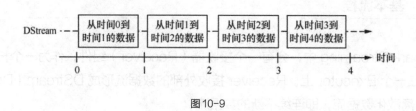

图 10-9

- 对 DStream 的操作也是以 RDD 为单位进行的。
- 底层 RDD 之间有依赖关系，DStream 之间也有依赖关系，RDD 具有容错性，那么 DStream 也具有容错性。
- Spark Streaming 将流式计算分解成多个 Spark 作业，对于每一时间段数据的处理都会经过 Spark DAG 分解以及 Spark 的任务集的调度过程。对于目前版本的 Spark Streaming 而言，其最小的 Batch Size 值为 0.5 ~ 5s。所以，Spark Streaming 能够满足流式准实时计算场景，对实时性要求非常高的场景则不太适合，如高频实时交易。

简单来说，DStream 就是对 RDD 的封装，我们对 DStream 进行操作，就是对 RDD 进行操作。DataFrame、Dataset、DStream 本质上都可以理解成 RDD，如图 10-10 所示。

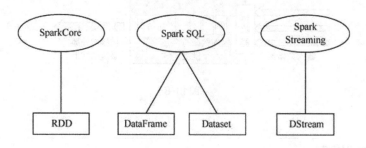

图 10-10

10.4 DStream 相关的 API

我们对 DStream 的操作，分为以下两种类型：Transformation 与 Output。下面对这两种类型的操作的 API 进行详细的介绍。

10.4.1 Transformation

Transformation 即无状态转换操作，每个批次的处理不依赖于之前批次的数据，如表 10-1 所示。

表 10-1

Transformation	含义
map(func)	对 DStream 中的各个元素进行 func 函数操作，然后返回一个新的 DStream
flatMap(func)	与 map 方法类似，只不过各个输入项可以被输出为零个或多个输出项
filter(func)	过滤出所有 func 函数的返回值为 true 的 DStream 元素并返回一个新的 DStream
union(otherDStream)	将源 DStream 和输入参数为 otherDStream 的元素合并，并返回一个新的 DStream
reduceByKey(func, [numTasks])	利用 func 函数对源 DStream 中的 key 进行聚合操作，然后返回新的 (K,V) 类型的 DStream
join(otherDStream, [numTasks])	输入 (K,V)、(K,W) 类型的 DStream，返回一个新的 (K, (V,W)) 类型的 DStream
transform(func)	通过 RDD-to-RDD 函数作用于 DStream 中的各个 RDD，可以是任意的 RDD 操作，从而返回一个新的 RDD

10.4.2 Output

Output 操作可以将 DStream 输出到外部的数据库或文件系统。当某个 Output 被调用时，Spark Streaming 程序才会开始真正的计算过程（与 RDD 的 Action 操作类似）。常见的 Output 操作如表 10-2 所示。

表 10-2

Output	含义
print	打印到控制台
saveAsTextFiles(prefix, [suffix])	保存流的内容为文本文件，文件名为"prefix-TIME_IN_MS[.suffix]"
saveAsObjectFiles(prefix,[suffix])	保存流的内容为 SequenceFile，文件名为"prefix-TIME_IN_MS[.suffix]"

续表

Output	含义
saveAsHadoopFiles(prefix,[suffix])	保存流的内容为 Hadoop 文件，文件名为"prefix-TIME_IN_MS[.suffix]"
foreachRDD(func)	对 DStream 里面的每个 RDD 执行 func 函数

10.5 Spark Streaming 原理总结

　　Spark Streaming 从 Kafka、Socket 或 HDFS 等获取数据，通过 Spark Streaming 引擎，不断地接收数据并按照时间间隔将其切分成多个 RDD 组成 DStream，之后对 DStream 进行一系列的 Transformation 和 Output 操作，最终将结果保存到目标服务器上。整个数据处理的流程如图 10-11 所示。

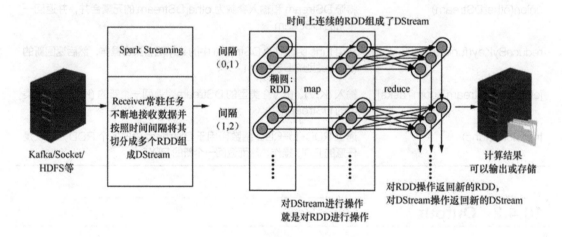

图 10-11

10.6 Spark Streaming 实战

Spark Streaming 第一个案例 WordCount

1. 需求与准备

　　生产者不断地生产单词，消费者每隔 5s 为一个批次，统计生产者在 5s 内所生产的单词重复出现的次数,需求概要架构如图10-12所示。

图10-12

首先,在Linux服务器上安装nc工具,nc是Netcat的简称,原本用来设置路由器,我们可以利用它向某个端口发送数据,安装命令如下。

```
yum install -y nc
```

其次,启动一个服务端并开放9999端口,等一下往这个端口发送数据,命令如下。

```
nc -lk 9999
```

最后,就可以发送数据了。

2. 代码演示

① 打开IDEA开发工具,并创建名为Spark Streaming的项目。

② 创建com.example.spark streaming包。

③ 创建NetworkWordCount.scala应用程序文件,代码如下。

```scala
package com.example.sparkstreaming
import org.apache.spark.SparkConf
import org.apache.spark.streaming.{Seconds, StreamingContext}
/*
  Spark Streaming应用程序,用来处理每5s内生产终端所
  发过来的单词重复出现的次数
 */
object NetworkWordCount{
  def main(args: Array[String]): Unit = {
        //配置信息
    val conf = new SparkConf().setMaster("local[2]").setAppName("NetworkWordCount")
      //创建StreamingContext对象,设置批处理的间隔时间为5s
val ssc = new StreamingContext(conf, Seconds(5))
//显示警告日志信息
ssc.sparkContext.setLogLevel("WARN")
        /*监听本地服务器上的9999端口,从中获取一行数据。ReceiverInputDStream就是由接
```

收到的所有数据组成的RDD，封装成了DStream，接下来对DStream进行操作就是对RDD进行操作*/
```scala
    val lines: ReceiverInputDStream = ssc.socketTextStream("master", 9999)
        //操作数据
        val words: DStream[String] = lines.flatMap(_.split(" "))
        //将单词进行映射，配对为(String, Int)
        val pairs: DStream[(String, Int)] = words.map(word=>(word,1))
        //按key分组，进行预统计
        val wordCounts: DStream[(String, Int)] = pairs.reduceByKey(_+_);
        //计算并将统计结果输出到控制台
        wordCounts.print()
        ssc.start()  //启动Spark Streaming应用程序
        ssc.awaitTermination()  //等待程序运行完后自动结束进程
    }
}
```

④ 打开一个终端，将 NetworkWordCount.scala 应用程序打包成 jar 包并提交到集群中运行，采用如下方法传数据。

```
 [ydh@master ~]$ nc -lk 9999
aa
aaa
hadoop hadoop hadoop
hbase hbase hbase
spring spring spring
hive hive
……
```

或者通过 SocketServer 发送数据到 Spark Streaming 的 NetworkWordCount.scala 应用程序中。需要注意的是，TcpServer 为服务端，即 TcpServer 应用程序要先启动，然后启动 Spark Streaming 的 NetworkWordCount.scala 应用程序，才能从 TcpServer 获取数据源。TcpServer 的代码如下。

```java
package com.example.sparkstreaming;
import java.io.PrintWriter;
import java.net.ServerSocket;
import java.net.Socket;

/*
 TcpServer就是以ServerSocket的程序代码的方式向Spark Streaming提供源
 源不断的数据的
 */
public class TcpServer {
```

```java
public static void main(String[] args)throws Exception{
    //创建ServerSocket并开放9999端口
    ServerSocket server = new ServerSocket(9999);
        //创建Socket，等待客户端的请求
    Socket socket = server.accept();
        //从Socket中获取输出对象并将其封装到字符输出流里
    PrintWriter pw = new PrintWriter(socket.getOutputStream());
        //准备数据
    String[] words = {"hadoop", "hbase", "hadoop", "hadoop", "hbase", "storm", "storm", "hbase", "hbase"};
        //持续不断地发送数据
    while(true){
        Thread.sleep(300);
        System.out.println("send words............");
        for(String word : words){
            pw.println(word);
            pw.flush();
        }
    }
}
```

⑤ 运行程序，观察 IDEA 控制台的输出。

在 IDEA 中运行 NetworkWordCount.scala 应用程序，控制台会根据程序中的设置每 5s 处理一次 DStream，输出如下结果，实现 Spark Streaming 的流式计算，即 Spark Streaming 每 5s 计算一次当前 5s 内产生的数据，然后将每个批次的数据输出。

```
20:18:20 INFO  DAGScheduler:54 - ResultStage 308 (print at AQuickExample.scala:17) finished in 0.017 s
20:18:20 INFO  DAGScheduler:54 - Job 154 finished: print at AQuickExample.scala:17, took 0.022634 s
-------------------------------------------
Time: 1559877500000 ms
-------------------------------------------
(aa,1)
(aaaa,1)
(hive,2)
(hadoop,3)
(spring,3)
(hbase,3)
```

10.7 updateStateByKey 算子

updateStateByKey 算子能够记录 DStream 的累计状态，允许我们维护任意状态，同时不断地用新信息更新它。该算子适用于数据的累加，包括一段时间内的数据的累加。需要注意的是，由于每个批次都输出自己批次的数据，这个时候若使用该算子，就会使各个批次之间产生联系。

10.7.1 WordCount 案例问题分析

在上面的 WordCount 案例中你有没有发现存在这样一个问题：每个批次的单词出现的次数都被正确地统计出来，但是结果不能累加！比如，第一个批次统计出 hadoop 这个单词出现了 3 次，第二个批次统计出 hadoop 这个单词出现了 9 次，这两个批次是独立的，但若我们想知道在第一个批次和第二个批次中单词 hadoop 总共出现了几次，在程序中是得不到的。这个时候若使用 updateStateByKey 算子就可以实现累加的功能。所以在实际应用和开发中，如果需要累加则需要使用 updateStateByKey(func) 来更新状态。

10.7.2 代码实现

我们对前面的 WordCount 案例进行改造，对每个批次单词出现的次数进行累加，代码如下：

```
package cn.itcast.streaming
import org.apache.spark.streaming.dstream.{DStream, ReceiverInputDStream}
import org.apache.spark.streaming.{Seconds, StreamingContext}
import org.apache.spark.{SparkConf, SparkContext}
object WordCountUpdateStateByKey {
  def main(args: Array[String]): Unit = {
    //1.创建StreamingContext
    //spark.master 应该被设置为 local[n], n > 1, 即至少分配一个CPU核资源
    val conf = new SparkConf().setAppName("wc").setMaster("local[*]")
    val sc = new SparkContext(conf)
    sc.setLogLevel("WARN")
    val ssc = new StreamingContext(sc,Seconds(5))//5表示5s内对数据进行切分，形成一个RDD
```

```
//检查点配置
//注意：我们在下面使用了updateStateByKey对当前数据和历史数据进行累加
//那么历史数据存在哪里？我们需要给它设置一个Checkpoint目录
ssc.checkpoint("./wc") //将处理过的数据存放到HDFS上
//2.监听Socket接收数据
/*ReceiverInputDStream就是由接收到的所有数据组成的RDD,封装成了DStream,接下来对
DStream进行操作就是对RDD进行操作*/
    val dataDStream: ReceiverInputDStream[String] = ssc.socketTextStream
("master",9999)
    //3.操作数据
    val wordDStream: DStream[String] = dataDStream.flatMap(_.split(" "))
    val wordAndOneDStream: DStream[(String, Int)] = wordDStream.map((_,1))
    //val wordAndCount: DStream[(String, Int)] = wordAndOneDStream.reduceByKey(_+_)
    //使用updateStateByKey对当前数据和历史数据进行累加
    val wordAndCount: DStream[(String, Int)]  =wordAndOneDStream.updateStateByKey
(updateFunc)
    wordAndCount.print()
    ssc.start()//开启
    ssc.awaitTermination()//等待优雅停止
  }
//currentValues：当前批次的value值，如1、1、1（以测试数据中的hadoop为例）
//historyValue：之前累计的历史值，第一次没有值则是0，第二次是3
//目标是把当前数据+历史数据返回作为新的结果（也是下次的历史数据）
  def updateFunc(currentValues:Seq[Int], historyValue:Option[Int] ):Op-
tion[Int] ={
    val result: Int = currentValues.sum + historyValue.getOrElse(0)
    Some(result)
  }
}
```

10.7.3 执行步骤

第一步，执行 nc -lk 9999。

第二步，执行以上代码。

第三步，不断地在第一步中输入不同的单词，如 hadoop spark sqoop hadoop spark hive hadoop。

第四步，观察 IDEA 控制台的输出，Spark Streaming 每隔 5s 计算一次当前批次内

的数据,然后将每个批次的结果数据累加输出。

10.8 reduceByKeyAndWindow 算子

reduceByKeyAndWindow(func,windowLength,slideInterval, [numTasks]) 基于滑动窗口对 (K,V) 类型的 DStream 中的值按 key 使用聚合函数 func 进行聚合操作,得到一个新的 DStream,其中 windowLength 为滑动窗口的长度。

10.8.1 图解 reduceByKeyAndWindow 算子

reduceByKeyAndWindow 算子即滑动窗口转换操作算子,是什么意思呢?继续用 WordCount 的案例来说明,比如我们想要每 10s 计算一下前 15s 的内容(每个词频统计的时间间隔是 5s),那我们可以想象到每 10s 计算出来的结果和前一次计算的结果其实中间有 5s 的值是重复的。

滑动窗口转换操作涉及的计算如下。我们可以事先设定一个滑动窗口的长度(也就是窗口的持续时间,例如 15s),并且设定滑动窗口的时间间隔(每隔多长时间执行一次计算,例如 10s),意思就是:每隔 10s 计算最近 15s 内的数据。这里出现了两个概念,即滑动窗口长度和滑动窗口的数据处理的时间间隔,如图 10-13 所示。

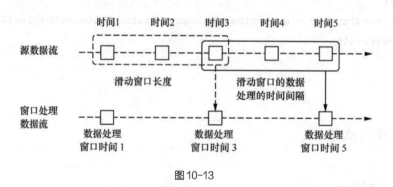

图 10-13

10.8.2 代码实现

```
package cn.stonesoup.streaming
import org.apache.spark.streaming.dstream.{DStream, ReceiverInputDStream}
```

```scala
import org.apache.spark.streaming.{Seconds, StreamingContext}
import org.apache.spark.{SparkConf, SparkContext}

object WordCountReduceByKeyAndWindow {
  def main(args: Array[String]): Unit = {
    //1.创建StreamingContext
    //spark.master应该被设置为local[n], n > 1, 即至少分配一个CPU核资源
    val conf = new SparkConf().setAppName("wc").setMaster("local[*]")
    val sc = new SparkContext(conf)
    sc.setLogLevel("WARN")
    val ssc = new StreamingContext(sc,Seconds(5))//5表示5s内对数据进行切分,形成一个RDD
    //2.监听Socket接收数据
    /*ReceiverInputDStream就是由接收到的所有数据组成的RDD,封装成了DStream,接下来对
DStream进行操作就是对RDD进行操作*/
    val dataDStream: ReceiverInputDStream[String] = ssc.socketTextStream("master",9999)
    //3.操作数据
    val wordDStream: DStream[String] = dataDStream.flatMap(_.split(" "))
    val wordAndOneDStream: DStream[(String, Int)] = wordDStream.map((_,1))
    //4.使用窗口函数进行词频统计
    //reduceFunc: (V, V) => V,集合函数
    //windowDuration : Duration,滑动窗口的长度/宽度
    //slideDuration : Duration,滑动窗口的时间间隔
    //注意: windowDuration值和slideDuration值必须是batchDuration值的倍数
    //windowDuration=slideDuration: 数据不会丢失也不会重复计算。开发中会使用
    //windowDuration>slideDuration: 数据会重复计算。开发中会使用
    //windowDuration<slideDuration: 数据会丢失
    //如果使用下面的代码
    //windowDuration=10
    //slideDuration=5
    //那么执行结果就是每隔10s计算最近15s内的数据
    //比如开发中让你统计最近1h内的数据,每隔1min计算一次,那么参数该如何设置呢
    //windowDuration=Minutes(60)
    //slideDuration=Minutes(1)
    val wordAndCount: DStream[(String, Int)] = wordAndOneDStream.((a:Int,b:Int)
=>a+b,Seconds(15),Seconds(10))
    wordAndCount.print()
    ssc.start()//开启
```

```
        ssc.awaitTermination()//等待程序运行结束
    }
}
```

10.8.3 执行步骤

第一步，执行 nc -lk 9999。

第二步，执行以上代码。

第三步，不断地在第一步中输入不同的单词，如 hadoop spark sqoop hadoop spark hive hadoop。

第四步，观察 IDEA 控制台的输出。

程序运行的现象：Spark Streaming 每隔 10s 计算一次最近 15s 内的数据，然后将结果数据输出。

10.9 统计一定时间内的热搜词

10.9.1 需求分析

我们来模拟一个百度热搜排行榜，一般在百度的搜索引擎的主界面的右下方会展示百度热搜，也就是此时此刻广大网民朋友搜索最多的内容，例如统计最近 10s 的热搜词的前 3 名，每隔 5s 计算一次。

想要实现上述功能，需要将滑动窗口的长度设置为 10s，即 windowDuration=10s；将滑动窗口的时间间隔设置为 5s，即 slideDuration=5s。

10.9.2 代码实现

```
package cn.stonesoup.streaming
import org.apache.spark.rdd.RDD
```

```scala
import org.apache.spark.streaming.dstream.{DStream, ReceiverInputDStream}
import org.apache.spark.streaming.{Seconds, StreamingContext}
import org.apache.spark.{SparkConf, SparkContext}
/**
 * 模拟百度热搜排行榜,统计最近10s的热搜词的前3名,每隔5s计算一次
 */
object WordCountTop3 {
  def main(args: Array[String]): Unit = {
    //1.创建StreamingContext
    //spark.master 应该被设置为local[n], n > 1,即至少分配一个CPU核资源
    val conf = new SparkConf().setAppName("wc").setMaster("local[*]")
    val sc = new SparkContext(conf)
    sc.setLogLevel("WARN")
    val ssc = new StreamingContext(sc,Seconds(5))//5表示5s内对数据进行切分,形成一个RDD
    //2.监听Socket接收数据
      /*ReceiverInputDStream就是由接收到的所有数据组成的RDD,封装成了DStream,接下来对DStream进行操作就是对RDD进行操作*/
    val dataDStream: ReceiverInputDStream[String] = ssc.socketTextStream("master", 9999)
    //3.操作数据
    val wordDStream: DStream[String] = dataDStream.flatMap(_.split(" "))
    val wordAndOneDStream: DStream[(String, Int)] = wordDStream.map((_,1))
    //4.使用窗口函数进行词频统计
    val wordAndCount: DStream[(String, Int)] = wordAndOneDStream.reduceByKey-
AndWindow((a:Int,b:Int)=>a+b,Seconds(10),Seconds(5))
    val sortedStream: DStream[(String, Int)] = wordAndCount.transform(rdd => {
        val sortedRDD: RDD[(String, Int)] = rdd.sortBy(_._2, false)  //逆序/降序
        println("===============top3==============")
        sortedRDD.take(3).foreach(println)
        println("===============top3==============")
        sortedRDD
    })
    //没有注册输出操作,所以没有执行任何操作
    sortedStream.print
    ssc.start()//开启
    ssc.awaitTermination()//等待优雅停止
  }
}
```

10.9.3 执行步骤

第一步，执行 nc -lk 9999。

第二步，执行以上代码。

第三步，不断地在第一步中输入不同的单词，如 hadoop spark sqoop hadoop spark hive hadoop。

第四步，观察 IDEA 控制台的输出，查看实际运行结果。

10.10 整合 Kafka

10.10.1 Kafka 基本概念

Kafka 是一种高吞吐量、分布式、快速、可扩展、分区和可复制的，基于发布与订阅模式的消息系统，是 Apache 项目集中的一个顶级项目。Kafka 使用 Scala 语言编写，目前已被广泛应用于各行业各类型的数据管道和消息系统中。Kafka 可以同时满足在线实时处理和批量离线处理。在大数据生态系统中，通常将 Kafka 作为数据交换枢纽，不同类型的系统（关系数据库、NoSQL 数据库、流处理系统、批处理系统等）可以统一接入 Kafka，实现和 Hadoop 各个组件之间的不同类型数据的实时高效交换。

10.10.2 Kafka 的特性

Kafka 的特性如下。

① 通过 $O(1)$ 的磁盘数据结构提供消息的持久化，这种结构对于即使 TB（10^{12}B）级的消息存储也能够保持长时间的稳定性能。

$O(1)$ 常数阶，是最低的时间复杂度，耗时或耗空间与输入数据大小无关，无论输入数据增大多少倍，耗时与耗空间都不变。哈希算法就是典型的时间复杂度为 $O(1)$ 的算法，无论数据规模多大，都可以在一次计算后找到目标（不考虑冲突的情况下）。

② 高吞吐量。即使是非常普通的硬件，Kafka 也可以支持每秒数十万的消息。

③ 支持通过 Kafka 服务器和客户机集群来对消息进行区分。

④ 支持 Hadoop 并行数据加载。

10.10.3 核心概念图解

一个典型的 Kafka 集群包含若干 Producer（生产者，可以是 Web 前端产生的 Page View，或者是服务器日志）、若干 Broker（协调者，Kafka 支持水平扩展，一般 Broker 数量越多，集群吞吐率越高）、若干 Consumer（消费者），以及一个 ZooKeeper 集群，如图 10-14 所示。

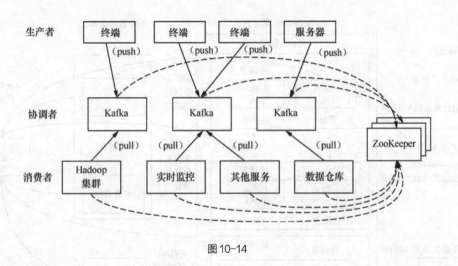

图 10-14

Kafka 通过 ZooKeeper 管理集群配置，选举 Leader（领导），以及在消费者组发生变化时进行负载均衡。

生产者使用 push 模式将消息发布到协调者，消费者使用 pull 模式从协调者处订阅并消费消息。

Broker 即消息的协调者，如安装 Kafka 的机器。

Producer 即消息的生产者，负责将数据写入 Broker 中（push 模式）。

Consumer 即消息的消费者，负责从 Kafka 中拉取数据（pull 模式）。旧版本的消费者需要依赖 ZooKeeper，新版本的不需要。

Topic 即主题，相当于数据的一个分类，不同主题存放不同业务的数据，可以用来区分业务。

Replication 即副本，数据保存多少份（为了保证数据不丢失）副本用以保证数据安全。

Partition 即分区，是一个物理的分区，一个分区就是一个文件，一个主题可以有 1～n 个分区，每个分区都有自己的副本，分区用以并发读写数据。

Consumer Group 即消费者组，一个主题可以有多个消费者或消费者组同时消费，多个消费者如果在一个消费者组中，那么它们不能重复消费数据。消费者组可以提高消费者消费速度、方便统一管理。

一个主题可以被多个消费者或者消费者组订阅，一个消费者或消费者组也可以订阅多个主题。需要注意的是：读数据只能从 Leader 读，写数据也只能往 Leader 写，Follower（跟从者）会从 Leader 那里同步数据过来作为副本，如图 10-15 所示。

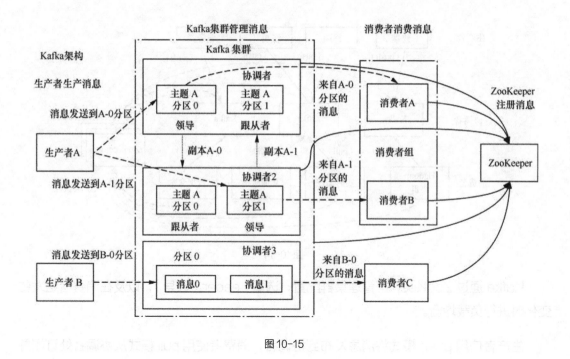

图 10-15

10.10.4　Kafka 集群部署

Kafka 通过 ZooKeeper 管理集群配置，选举领导，以及在消费者组发生变化时进行负载均衡。所以部署 Kafka 集群之前需先部署 ZooKeeper 集群。

① 下载 Kafka 安装包，登录 Kafka 官方网站即可下载。

② 上传安装包并解压，命令如下。

```
tar -zxvf kafka_2.11-1.0.0.tgz -C /opt/soft/
cd /opt/soft/
mv kafka_2.11-1.0.0 kafka
```

③ 配置环境变量。

```
vim /etc/profile
#KAFKA_HOME
export KAFKA_HOME=/opt/soft/kafka
export PATH=$PATH:$KAFKA_HOME/bin
source /etc/profile
```

④ 分别向 master、slave、slave1 分发安装包。

```
#分发安装包
scp -r /opt/soft/kafka   master:/opt/soft
scp -r /opt/soft/kafka   slave:/opt/soft
scp -r /opt/soft/kafka   slave1:/opt/soft
#分发环境变量文件
scp /etc/profile master:/etc/profile
scp /etc/profile slave:/etc/profile
scp /etc/profile slave1:/etc/profile
#使环境变量文件生效
source /etc/profile
```

⑤ 修改 Kafka 配置文件。

```
cp/opt/soft/kafka/config/server.properties /opt/soft/kafka/config/server.properties.bak
vim /opt/soft/kafka/config/server.properties
```

在修改 Kafka 配置文件之前，我们对 server.properties 配置文件中的每一个配置项进行说明，具体如下。

```
# 设置Kafka 节点唯一ID
broker.id=0
# 开启删除Kafka 主题属性
delete.topic.enable=true
# 非SASL 模式配置Kafka集群
listeners=PLAINTXT://master:9092
# 设置网络请求处理线程数
num.network.threads=10
# 设置磁盘I/O 请求线程数
num.io.threads=20
```

```
# 设置发送buffer字节数
socket.send.buffer.bytes=1024000
# 设置收到buffer字节数
socket.receive.buffer.bytes=1024000
# 设置最大请求字节数
socket.request.max.bytes=1048576000
# 设置消息记录存储路径
log.dirs=/opt/data/kafka/kafka-logs
# 设置Kafka的主题分区数
num.partitions=2
# 设置主题保留时间
log.retention.hours=168
# 设置ZooKeeper的连接地址
zookeeper.connect=master:2181,slave:2181,slave1:2181
# 设置ZooKeeper连接起始时间
zookeeper.connection.timeout.ms=60000
```

在第一台机器 master 上修改 Kafka 配置文件 server.properties，命令如下。

```
vim /opt/soft/kafka/config/server.properties
```

首先删除 server.properties 文件中已有的所有内容之后，再在 server.properties 中输入以下内容。

```
broker.id=0
num.network.threads=3
num.io.threads=8
socket.send.buffer.bytes=102400
socket.receive.buffer.bytes=102400
socket.request.max.bytes=104857600
log.dirs=/opt/data/kafka/kafka-logs
num.partitions=2
num.recovery.threads.per.data.dir=1
offsets.topic.replication.factor=1
transaction.state.log.replication.factor=1
transaction.state.log.min.isr=1
log.flush.interval.messages=10000
log.flush.interval.ms=1000
log.retention.hours=168
log.segment.bytes=1073741824
log.retention.check.interval.ms=300000
zookeeper.connect=master:2181,slave:2181,slave1:2181
zookeeper.connection.timeout.ms=6000
```

```
group.initial.rebalance.delay.ms=0
delete.topic.enable=true
host.name=master
```

在第二台机器 slave 上修改 Kafka 配置文件 server.properties,命令如下。

```
vim /opt/soft/kafka/config/server.properties
```

在 server.properties 中输入以下内容。

```
broker.id=1
num.network.threads=3
num.io.threads=8
socket.send.buffer.bytes=102400
socket.receive.buffer.bytes=102400
socket.request.max.bytes=104857600
log.dirs=/opt/data/kafka/kafka-logs
num.partitions=2
num.recovery.threads.per.data.dir=1
offsets.topic.replication.factor=1
transaction.state.log.replication.factor=1
transaction.state.log.min.isr=1
log.flush.interval.messages=10000
log.flush.interval.ms=1000
log.retention.hours=168
log.segment.bytes=1073741824
log.retention.check.interval.ms=300000
zookeeper.connect=master:2181,slave:2181,slave1:2181
zookeeper.connection.timeout.ms=6000
group.initial.rebalance.delay.ms=0
delete.topic.enable=true
host.name=slave
```

在第三台机器 slave1 上修改 Kafka 配置文件 server.properties,命令如下。

```
vim /opt/soft/kafka/config/server.properties
```

在 server.properties 中输入以下内容。

```
broker.id=2
num.network.threads=3
num.io.threads=8
socket.send.buffer.bytes=102400
socket.receive.buffer.bytes=102400
```

```
socket.request.max.bytes=104857600
log.dirs=/opt/data/kafka/kafka-logs
num.partitions=2
num.recovery.threads.per.data.dir=1
offsets.topic.replication.factor=1
transaction.state.log.replication.factor=1
transaction.state.log.min.isr=1
log.flush.interval.messages=10000
log.flush.interval.ms=1000
log.retention.hours=168
log.segment.bytes=1073741824
log.retention.check.interval.ms=300000
zookeeper.connect=master:2181,slave:2181,slave1:2181
zookeeper.connection.timeout.ms=6000
group.initial.rebalance.delay.ms=0
delete.topic.enable=true
host.name=slave1
```

1. 启动 Kafka 集群

先启动 ZooKeeper 集群。

然后在 3 台机器 master、slave 和 slave1 上分别启动 Kafka 集群，启动命令如下。

```
/opt/soft/kafka/bin/kafka-server-start.sh -daemon config/server.properties
```

创建一个 test 的主题。

```
/opt/soft/kafka/bin/kafka-topics.sh --create --zookeeper master:2181 --replication-factor 1 --partitions 1 --topic test
```

在一个终端上启动一个生产者。

```
/opt/soft/kafka/bin/kafka-console-producer.sh --broker-list master:9092 --topic test
```

然后输入下面的信息。

明天会更好。
Hello world!

之后在另一个终端上启动一个消费者。

```
bin/kafka-console-consumer.sh --zookeeper hadoop1:2181 --topic test --from-beginning
```

此时会发现屏幕上有同样的信息输出，证明 Kafka 集群已经搭建成功。输出的内容如下。

> 明天会更好。
> Hello world!

2. 关闭 Kafka 集群

关闭 Kafka 集群命令如下。

`/opt/soft/kafka/bin/kafka-server-stop.sh stop`

或者使用：

`kafka-cmd.sh stop`

10.10.5 Kafka 常用命令

1. 启动 Kafka

`/export/servers/kafka/bin/kafka-server-start.sh -daemon /export/servers/kafka/config/server.properties`

2. 停止 Kafka

`/export/servers/kafka/bin/kafka-server-stop.sh`

3. 查看主题信息

`/export/servers/kafka/bin/kafka-topics.sh --list --zookeeper master:2181`

4. 创建主题

`/export/servers/kafka/bin/kafka-topics.sh --create --zookeeper master:2181 --replication-factor 3 --partitions 3 --topic test`

5. 查看某个主题信息

`/export/servers/kafka/bin/kafka-topics.sh --describe --zookeeper master:2181 --topic test`

6. 删除主题

```
/export/servers/kafka/bin/kafka-topics.sh --zookeeper master:2181 --delete
--topic test
```

7. 启动生产者（控制台的生产者一般用于测试）

```
/export/servers/kafka/bin/kafka-console-producer.sh --broker-list master:9092
--topic spark_kafka
```

8. 启动消费者（控制台的消费者一般用于测试）

```
/export/servers/kafka/bin/kafka-console-consumer.sh --zookeeper master:2181
--topic spark_kafka--from-beginning
```

9. 消费者连接到协调者的地址

```
/export/servers/kafka/bin/kafka-console-consumer.sh --bootstrap-server master:
9092,slave:9092,slave1:9092 --topic spark_kafka --from-beginning
```

Spark Streaming 整合 Kafka 两种模式说明在实际项目开发中我们经常会利用 Spark Streaming 实时地读取 Kafka 中的数据，然后进行计算处理。在 Spark 1.3 后，KafkaUtils 工具类里面提供了两种创建 DStream 的方法：Receiver 接收方式和 Direct 直连方式。

10.10.6 Receiver 接收方式

通过 KafkaUtils.createDstream 方法可以从 Kafka 中得到 DStream，这种方式在实际项目开发中已经不再使用，这里作为扩展知识了解即可。

Receiver 作为常驻的任务运行在 Executor 中等待数据，但是一个 Receiver 效率很低，需要开启多个 Receiver，再手动合并（union）数据并进行处理，这将会变得非常麻烦。一旦运行 Receiver 的那台机器挂了，可能会丢失数据，所以需要开启 WAL（Write-Ahead Logging，预写日志）来保证数据安全，这样一来效率又会降低。

需要注意的是，Receiver 接收方式是通过 ZooKeeper 来连接到 Kafka 队列的，从而调用 Kafka 高阶 API，偏移量（offset）存储在 ZooKeeper 中，由 Receiver 维护，Spark 在消费的时候为了保证数据不丢失，也会在 Checkpoint 中存储一份 offset，因此可能会出现数据不一致的情况。所以不管从何种角度来说，Receiver 接收方式都不适合在实际项目开发中使用。

10.10.7 Direct 直连方式

通过 KafkaUtils.createDirectStream 方法可以从 Kafka 中得到 DStream，这种方式在实际项目开发中经常使用，需要重点掌握。

Direct 直连方式是直接连接 Kafka 分区来获取数据的，从每个分区直接读取数据大大提高了并行能力。Direct 直连方式调用 Kafka 低阶 API（底层 API），自己存储和维护 offset，默认由 Spark 维护在 Checkpoint 中，消除了与 ZooKeeper 不一致的情况。当然也可以自己手动维护，把 offset 存储在 MySQL、Redis 中。所以 Direct 直连方式可以在开发中使用，且借助 Direct 直连方式的特点 + 手动操作可以保证数据只被处理一次。

Receiver 接收方式与 Direct 直连方式的消息语义定义如表 10-3 所示。

表 10-3

实现方式	消息语义定义	存在的问题
Receiver	at most once（最多被处理一次）	会丢失数据
Receiver+WAL	at least once（最少被处理一次）	不会丢失数据，但可能会重复消费，且效率低
Direct+ 手动操作	exactly once（只被处理一次/精准一次）	不会丢失数据，也不会重复消费，且效率高

需要注意的是，在实际项目开发中 Spark Streaming 和 Kafka 集成时有两个版本可以选择：0.8 及 0.10+。其中 0.8 版本有 Receiver 和 Direct 方式，但是 0.8 版本生产环境问题较多，在 Spark 2.3 之后就已经不再支持 0.8 版本了。0.10 版本以后只保留了 Direct 直连方式（因为通过前面的讲解我们已经知道 Reveiver 接收方式不适合生产环境），并且 0.10 版本的 API 有变化，功能更加强大。

我们在实际项目开发时直接使用 0.10 版本中的 Direct 直连方式，但是关于 Receiver 接收方式和 Direct 直连方式的区别也能够分析出来。

10.10.8 spark-streaming-kafka-0-8 版本

1. Receiver 接收方式

KafkaUtils.createDstream 使用 Receiver 来接收数据，利用的是 Kafka 高阶 API， offset 由 Receiver 在 ZooKeeper 中维护，对于所有 Receiver 接收到的数据将会保存在 Executor 中，然后通过 Spark Streaming 启动作业来处理这些数据，默认会丢失，可启用 WAL，它将接收到的数据同步保存到分布式文件系统（如 HDFS）上，以保证数据在出错的情况下可以恢复，如图 10-16 所示。尽管这种方式配合 WAL 机制可以保证数据零丢失的高可靠性，但是启用了 WAL 效率会较低，且无法保证数据仅被处理一次，可能会被处理两次。因为 Spark 和 ZooKeeper 之间可能是不同步的。需要注意的是，官方现在已经不推荐这种整合方式。

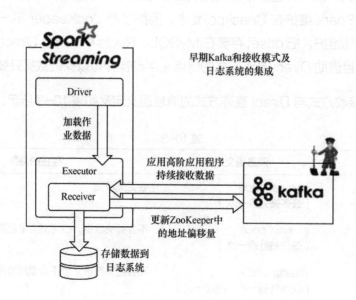

图 10-16

（1）做好一些基本的准备工作。

① 启动 ZooKeeper 集群。

```
zkServer.sh start
```

② 启动 Kafka 集群。

```
kafka-server-start.sh /opt/soft/kafka/config/server.properties
```

③ 创建主题。

```
kafka-topics.sh --create --zookeeper master:2181 --replication-factor 1 --partitions 3 --topic spark_kafka
```

④ 通过 Shell 命令向主题发送消息。

```
kafka-console-producer.sh --broker-list master:9092 --topic spark_kafka
hadoop spark sqoop hadoop spark hive hadoop
```

⑤ 添加 Kafka 的 POM 依赖。

```xml
<dependency>
    <groupId>org.apache.spark</groupId>
    <artifactId>spark-streaming-kafka-0-8_2.11</artifactId>
    <version>2.2.0</version>
</dependency>
```

(2) 通过 API 设置 Receiver 的个数,用于并发地从 Kafka 的主题中获取数据。

通过 Receiver 获取 Kafka 的主题中的数据,可以并行运行更多的 Receiver 来获取 Kafka 的主题中的数据,这里我们将其个数设置为 3,如下。

```
val receiverDStream: immutable.IndexedSeq[ReceiverInputDStream[(String, String)]] = (1 to 3).map(x => {
    val stream: ReceiverInputDStream[(String, String)] = KafkaUtils.createStream(ssc, zkQuorum, groupId, topics)
    stream })
```

如果启用了 WAL(spark.streaming.receiver.writeAheadLog.enable=true),可以设置存储级别(默认为 StorageLevel.MEMORY_AND_DISK_SER_2)。

(3) 代码实现。

```scala
package cn.soupstone.streaming
import org.apache.spark.streaming.dstream.{DStream, ReceiverInputDStream}
import org.apache.spark.streaming.kafka.KafkaUtils
import org.apache.spark.streaming.{Seconds, StreamingContext}
import org.apache.spark.{SparkConf, SparkContext}
import scala.collection.immutable
object SparkKafka {
  def main(args: Array[String]): Unit = {
    //1.创建StreamingContext
    val config: SparkConf =
new SparkConf().setAppName("SparkStream").setMaster("local[*]")
      .set("spark.streaming.receiver.writeAheadLog.enable", "true")
//开启WAL,保证数据源端可靠性
    val sc = new SparkContext(config)
    sc.setLogLevel("WARN")
    val ssc = new StreamingContext(sc,Seconds(5))
    ssc.checkpoint("./kafka")
    //2.准备配置参数
    val zkQuorum = "master:2181,slave:2181,slave1:2181"
    val groupId = "spark"
    val topics = Map("spark_kafka" -> 2)//2表示每一个主题对应的分区都采用2个线程去消费
//Spark Streaming的RDD分区和Kafka的主题分区不一样,增加消费线程数,并不增加Spark的并行处
```

理数据数量

```
/*3.通过Receiver获取Kafka主题中的数据,可以并行运行更多的Receiver来获取Kafka主题中的数据,
这里将其个数设置为3*/
    val receiverDStream: immutable.IndexedSeq[ReceiverInputDStream[(String,
String)]] = (1 to 3).map(x => {
        val stream: ReceiverInputDStream[(String, String)] = KafkaUtils.createStream
(ssc, zkQuorum, groupId, topics)
        stream
    })
    //4.使用union方法,将所有Receiver产生的DStream进行合并
    val allDStream: DStream[(String, String)] = ssc.union(receiverDStream)
    //5.获取主题中的数据(String, String),第一个String表示主题的名称,第二个String表
示主题中的数据
    val data: DStream[String] = allDStream.map(_._2)
    //6.进行词频统计
    val words: DStream[String] = data.flatMap(_.split(" "))
    val wordAndOne: DStream[(String, Int)] = words.map((_, 1))
    val result: DStream[(String, Int)] = wordAndOne.reduceByKey(_ + _)
    result.print()
    ssc.start()
    ssc.awaitTermination()
  }
}
```

2. Direct 直连方式

Direct 直连方式会定期地从 Kafka 的主题对应的分区中查询最新的 offset,再根据 offset 范围在每个批次里面处理数据,Spark 通过调用 Kafka 简单的消费者 API 读取一定范围的数据,如图 10-17 所示。

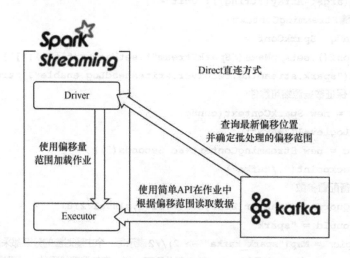

图 10-17

① Direct 直连方式的缺点是无法使用基于 ZooKeeper 的 Kafka 监控工具。

② Direct 直连方式相比 Receiver 接收方式有以下几个优点。

● 简化并行。

不需要创建多个 Kafka 输入流,然后合并它们,Spark Streaming 将会创建和 Kafka 分区数一样的 RDD 的分区数,而且会从 Kafka 中并行读取数据,Spark 中 RDD 的分区数和 Kafka 中的分区数是一一对应的。

● 高效。

Receiver 可实现数据的零丢失是因为将数据预先保存在 WAL 中,会复制一遍数据,这导致数据被复制两次,第一次是被 Kafka 复制,第二次是写到 WAL 中。而 Direct 直连方式不使用 WAL,消除了这个问题。

Receiver 读取 Kafka 数据是通过 Kafka 高阶 API 把 offset 写入 ZooKeeper 中,虽然这种方式可以将数据保存在 WAL 中,保证数据不丢失,但是可能会因为 Spark Streaming 和 ZooKeeper 中保存的 offset 不一致而导致数据被消费多次。

Direct 的 EOS(Exactly-once-semantics)通过实现 Kafka 低阶 API,offset 仅被 Spark Streaming 保存在 Checkpoint 中,消除了 ZooKeeper 和 Spark Streaming 的 offset 不一致的问题。

● API。

Spark Streaming 对接 Kafka,用 Direct 直连方式消费数据的方法,其有 3 个参数:SSC 指 Spark Streaming Context 对象;kafkaParams 指 Kafka 相关配置信息;topic 指主题信息,需要消费数据时使用。

```
KafkaUtils.createDirectStream[String, String, StringDecoder, StringDecoder]
(ssc, kafkaParams, topics)
```

③ 代码实现。

```
package cn.itcast.streaming
import kafka.serializer.StringDecoder
import org.apache.spark.streaming.dstream.{DStream, InputDStream}
import org.apache.spark.streaming.kafka.KafkaUtils
import org.apache.spark.streaming.{Seconds, StreamingContext}
import org.apache.spark.{SparkConf, SparkContext}
```

```scala
//Direct直连方式从Kafka中获取数据
object SparkKafka2 {
  def main(args: Array[String]): Unit = {
    //1.创建StreamingContext
    val config: SparkConf =
  new SparkConf().setAppName("SparkStream").setMaster("local[*]")
    val sc = new SparkContext(config)
    sc.setLogLevel("WARN")
    val ssc = new StreamingContext(sc,Seconds(5))
    ssc.checkpoint("./kafka")

    //2.准备配置参数
    val kafkaParams = Map("metadata.broker.list" -> "master:9092,master:9092,master:9092", "group.id" -> "spark")
    val topics = Set("spark_kafka")
    val allDStream: InputDStream[(String, String)] = KafkaUtils.createDirectStream[String, String, StringDecoder, StringDecoder](ssc, kafkaParams, topics)
    //3.获取主题中的数据
    val data: DStream[String] = allDStream.map(_._2)
    //WordCount
    val words: DStream[String] = data.flatMap(_.split(" "))
    val wordAndOne: DStream[(String, Int)] = words.map((_, 1))
    val result: DStream[(String, Int)] = wordAndOne.reduceByKey(_ + _)
    result.print()
    ssc.start()
    ssc.awaitTermination()
  }
}
```

10.10.9 spark-streaming-kafka-0-10 版本

在 spark-streaming-kafka-0-10 版本中,API 有一定的变化,操作更加灵活。这个版本在实际项目开发中经常使用,需要重点掌握。

① 在项目中通过 pom.xml 引入 spark-streaming-kafka-0-10 版本的依赖包。

```xml
<!--<dependency>
    <groupId>org.apache.spark</groupId>
    <artifactId>spark-streaming-kafka-0-8_2.11</artifactId>
    <version>${spark.version}</version>
</dependency>-->
<dependency>
    <groupId>org.apache.spark</groupId>
    <artifactId>spark-streaming-kafka-0-10_2.11</artifactId>
    <version>${spark.version}</version>
</dependency>
```

② 在 Kafka 中创建主题，命令如下。

```
/opt/soft/kafka/bin/kafka-topics.sh --create --zookeeper master:2181 --replication-factor 3 --partitions 3 --topic spark_kafka
```

③ 启动生产者，准备向 Kafka 集群中生产数据，命令如下。

```
/opt/soft/kafka/bin/kafka-console-producer.sh --broker-list master:9092,slave:9092,slave1:9092 --topic spark_kafka
```

④ 代码实现。

```scala
package cn.itcast.streaming

import org.apache.kafka.clients.consumer.ConsumerRecord
import org.apache.kafka.common.serialization.StringDeserializer
import org.apache.spark.streaming.dstream.{DStream, InputDStream}
import org.apache.spark.streaming.kafka010.{ConsumerStrategies, KafkaUtils, LocationStrategies}
import org.apache.spark.streaming.{Seconds, StreamingContext}
import org.apache.spark.{SparkConf, SparkContext}
//Direct直连方式从Kafka中消费数据
object SparkKafkaDemo {
  def main(args: Array[String]): Unit = {
    //1.创建StreamingContext
    //spark.master 应该被设置为 local[n], n > 1, 即至少分配一个CPU核资源
    val conf = new SparkConf().setAppName("wc").setMaster("local[*]")
    val sc = new SparkContext(conf)
    sc.setLogLevel("WARN")
    val ssc = new StreamingContext(sc,Seconds(5))//5表示5s内对数据进行切分，形成一个RDD
```

```scala
        //准备连接Kafka的参数
        val kafkaParams = Map[String, Object](
          "bootstrap.servers" -> "master:9092,slave:9092,slave1:9092",
          "key.deserializer" -> classOf[StringDeserializer],
          "value.deserializer" -> classOf[StringDeserializer],
          "group.id" -> "SparkKafkaDemo",
          /*earliest：当各分区下有已提交的offset时，从提交的offset开始消费；无提交的offset
时，从头开始消费。latest：当各分区下有已提交的offset时，从提交的offset开始消费；无提交的
offset时，消费新产生的该分区下的数据。none：当各分区都存在已提交的offset时，从offset开始
消费；只要有一个分区不存在已提交的offset，就抛出异常。这里配置latest，自动重置offset为最新的
offset，即如果有offset则从offset开始消费，没有offset则从新来的数据开始消费*/
          "auto.offset.reset" -> "latest",
          //false表示关闭自动提交，由Spark帮你提交到Checkpoint或程序员手动维护
          "enable.auto.commit" -> (false: java.lang.Boolean)
        )
    val topics = Array("spark_kafka")
        //2.使用KafkaUtils连接Kafka获取数据
        val recordDStream: InputDStream[ConsumerRecord[String, String]] = KafkaUtils.
createDirectStream[String, String](ssc,
          LocationStrategies.PreferConsistent,
          //位置策略，源码强烈推荐使用该策略，会让Spark的Executor和Kafka的Broker均匀对应
          ConsumerStrategies.Subscribe[String, String](topics, kafkaParams))
          //消费策略，源码强烈推荐使用该策略
        //3.操作数据
        val lineDStream: DStream[String] = recordDStream.map(_.value())
        //_指的是一行消费数据
        val wrodDStream: DStream[String] = lineDStream.flatMap(_.split(" "))
        //_指的是发过来的value，即一行数据
        val wordAndOneDStream: DStream[(String, Int)] = wrodDStream.map((_,1))
        val result: DStream[(String, Int)] = wordAndOneDStream.reduceByKey(_+_)
        result.print()
        ssc.start()//开启
        ssc.awaitTermination()//等待优雅停止
      }
    }
```

10.11 本章总结

　　本章主要介绍了 Spark Streaming 框架的基本概念,它是 Spark 用于处理实时数据的框架。随着大数据应用的发展,人们对实时数据的处理要求越来越多,Spark 巧妙地通过批次处理的方式实现实时数据的处理。接着讲述了 Spark Streaming 实战,体验 Spark Streaming 的流式数据处理。最后介绍了 Spark Streaming 与 Kafka 框架的整合开发。

10.12 本章习题

1. 完成 Spark Streaming 实战的代码开发。

2. 完成 Spark Streaming 与 Kafka 框架的整合开发。

第 11 章 Spark 综合项目实战

11.1 网站运营指标统计项目

网站运营指标统计项目主要功能是分析网站的历史或实时数据，从而对该网站运营指标进行评估。

例如，我们想知道网站的用户数、每日的浏览量、重复访问者的数量等，借此探索用户行为指标，以便了解用户如何来到网站、用户在网站上停留了多长时间、用户访问了哪些页面、用户使用的搜索引擎及关键词等信息。如此一来我们将会更好地维护网站用户，开发更适合用户使用的功能，吸引更多的用户，提供更优质的产品和服务，使得网站的效益进一步提升。

11.1.1 需求分析

1. 统计网站 PV

网站 PV（Page View）是指网站页面浏览量，也叫单击量，通常是衡量一个网络新闻频道或网站甚至一条网络新闻的主要指标。严格来说，网站 PV 就是一个访问者在 24h（0 时到 24 时）内浏览网站的页面数。需要注意的是，同一个人浏览网站的同一个页面，不重复计算 PV，就算他点击浏览了 50 次，也算一次。也就是说，PV 就是一个访问者打开网站的页面数。

网站 PV 之于网站就像收视率之于电视，从某种程度上已成为企业衡量商业网站表现的重要的尺度。

2. 统计网站 UV

UV（Unique Visitor）是指独立用户（也可以指通过互联网浏览网页的自然人），即一个用户算一次访问，可以使用 IP 地址 /SessionID 来计算。

同一天内，UV 只记录第一次访问网站的独立 IP 地址，若此 IP 地址再次访问该网站则不计数。确切来说，是指访问某个网站或单击某条新闻的不同 IP 地址的人数。

虽然 IP 地址与 UV 类似，但现在 IP 地址已经很难真实反映网站流量的实际情况，所以引入了更加精确的 UV 概念。

3. 统计网站用户来源前几项

我们可以根据网站用户来源的 RefURL 信息来分析网站用户都是通过什么渠道进入该网站的，常见的来源有搜索引擎、广告、咨询、媒体等，我们在本项目中可以统计出用户来源的前几项。

11.1.2 数据分析

查看 access.log 数据文件，其中记录了某网站用户某一天的访问数据信息，包含用户 IP 地址、用户访问网站时间、用户通过请求 URL 连接以及用户所使用的浏览器信息等，如图 11-1 所示。

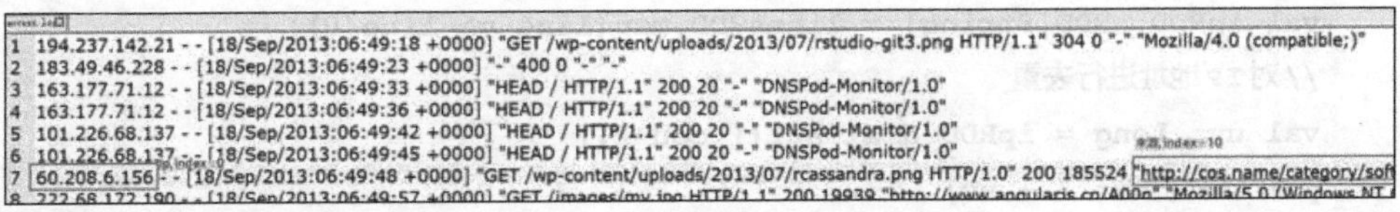
1 194.237.142.21 - - [18/Sep/2013:06:49:18 +0000] "GET /wp-content/uploads/2013/07/rstudio-git3.png HTTP/1.1" 304 0 "-" "Mozilla/4.0 (compatible;)"
2 183.49.46.228 - - [18/Sep/2013:06:49:23 +0000] "-" 400 0 "-" "-"
3 163.177.71.12 - - [18/Sep/2013:06:49:33 +0000] "HEAD / HTTP/1.1" 200 20 "-" "DNSPod-Monitor/1.0"
4 163.177.71.12 - - [18/Sep/2013:06:49:36 +0000] "HEAD / HTTP/1.1" 200 20 "-" "DNSPod-Monitor/1.0"
5 101.226.68.137 - - [18/Sep/2013:06:49:42 +0000] "HEAD / HTTP/1.1" 200 20 "-" "DNSPod-Monitor/1.0"
6 101.226.68.137 - - [18/Sep/2013:06:49:45 +0000] "HEAD / HTTP/1.1" 200 20 "-" "DNSPod-Monitor/1.0"
7 60.208.6.156 - - [18/Sep/2013:06:49:48 +0000] "GET /wp-content/uploads/2013/07/rcassandra.png HTTP/1.0" 200 185524 "http://cos.name/category/soft

图 11-1

11.1.3 代码实现

```
package cn.stone.core
import org.apache.spark.rdd.RDD
```

```scala
import org.apache.spark.storage.StorageLevel
import org.apache.spark.{SparkConf, SparkContext}

object WebLog {
  def main(args: Array[String]): Unit = {
    //1.创建SparkContext对象
    val conf: SparkConf = new SparkConf().setAppName("wc").setMaster("local[*]")
    val sc: SparkContext = new SparkContext(conf)
    sc.setLogLevel("WARN")

    //2.加载数据
    val fileRDD: RDD[String] = sc.textFile("D:\\data\\access.log")
    //3.处理数据,每一行按空格切分
    val linesRDD: RDD[Array[String]] = fileRDD.map(_.split(" "))
    //频繁使用的RDD可以进行缓存
    linesRDD.persist(StorageLevel.MEMORY_AND_DISK)
    //设置Checkpoint目录
    sc.setCheckpointDir("./ckp")//实际开发中写HDFS
    //将RDD进行Checkpoint
    linesRDD.checkpoint()

    //4.统计指标
    //统计网站PV
    val pv: Long = linesRDD.count()
    println("pv: " + pv)
    val pvAndCount: RDD[(String, Int)] = linesRDD.map(line => ("pv",1)).reduceByKey(_+_)
    pvAndCount.collect().foreach(println)

    //统计网站UV
    //取出IP地址
    val ipRDD: RDD[String] = linesRDD.map(line => line(0))
    //对IP地址进行去重
    val uv: Long = ipRDD.distinct().count()
    println("uv: " + uv)

    //统计网站用户来源前几项
    //统计RefURL,表示来自哪里
    //数据预处理
    val filteredLines: RDD[Array[String]] = linesRDD.filter(_.length > 10)
    //取出RefURL
    val refurlRDD: RDD[String] = filteredLines.map(_(10))
    //每个RefURL记为1
```

```
        val refurlAndOne: RDD[(String, Int)] = refurlRDD.map((_,1))
        //按照key聚合
        val refurlAndCount: RDD[(String, Int)] = refurlAndOne.reduceByKey(_+_)
        //val top5: Array[(String, Int)] = refurlAndCount.top(5)
        //注意：top默认按照key排序，我们需要按照value排序
        //并且top是把数据拉回Driver再排序
        val result = refurlAndCount.sortBy(_._2,false)//按照value降序排列
        val top5: Array[(String, Int)] = result.take(5)
        top5.foreach(println)
    sc.stop()
    }
}
```

11.2 热力图分析项目

11.2.1 需求分析

我们在互联网中经常见到城市热力图。例如百度统计工具会统计热门旅游城市、热门报考学校、热门餐饮店等，并将这样的信息显示在热力图中。因此，我们需要通过日志信息（运营商或者网站自己生成）和城市IP地址段信息来判断用户的IP地址段，统计热点经纬度、热门城市等指标。接下来我们使用Spark技术来实现上述功能。

11.2.2 数据分析

数据分为两种。20190121000132.394251.http.format 是某城市用户访问互联网的基础日志数据信息，其中包含用户的访问日期、IP地址信息、网址、浏览器信息以及用户源信息等。

数据文件 ip.txt 中 IP 地址规则数据信息包含开始 IP 地址、结束 IP 地址、开始数字、结束数字、洲、国家、省、市、区、运营商、行政区域、英文、代码、经纬度等。

IP 地址规则数据在使用的时候要知道它的每一项的具体含义，如图 11-2 所示。

开始IP地址	结束IP地址	开始数字	结束数字	洲	国家	省	市	区	运营商	行政区域	英文	代码	经纬度
1.0.1.0	1.0.3.255	16777472	16778239	亚洲	中国	福建	福州		电信	350100	China	CN	119.306239|26.075302

图 11-2

注意，在本项目的 20190121000132.394251.http.format 中，只需要关心其中的 IP 地址信息就可以了，其他的数据项暂时用不到。

11.2.3 项目开发

1. RDD 实现方式

① 创建 SparkContext。

② 加载 IP 地址规则文件。

③ 获取 IP 地址起始范围和结束范围、城市信息、经度、纬度。

④ 加载日志文件。

⑤ 将日志中的 IP 地址分割出来。

⑥ 将 IP 地址转为数字，并使用二分查找法去 ipRules 中查找出相应的城市信息，记为 ((城市,经度,纬度),1)。

⑦ 将 ((城市,经度,纬度),1) 进行聚合得出统计结果。

⑧ 将结果数据写入 MySQL。

准备数据库表，在 MySQL 数据库中创建即可。

```
CREATE TABLE 'iplocaltion' (
  'id' bigint(20) NOT NULL AUTO_INCREMENT,
  'city' varchar(255) DEFAULT NULL,
  'longitude' varchar(255) DEFAULT NULL,
  'latitude' varchar(255) DEFAULT NULL,
  'total_count' bigint(20) DEFAULT NULL,
  PRIMARY KEY ('id')
) ENGINE=InnoDB DEFAULT CHARSET=utf8;
```

代码实现如下。

```
package cn.itcast.ip
import java.sql.{Connection, DriverManager, PreparedStatement}

import org.apache.spark.broadcast.Broadcast
import org.apache.spark.rdd.RDD
```

```scala
import org.apache.spark.{SparkConf, SparkContext}

object IPMap {
  def main(args: Array[String]): Unit = {
    //1.创建SparkContext
    val conf: SparkConf = new SparkConf().setAppName("ip").setMaster("local[*]")
    val sc = new SparkContext(conf)
    sc.setLogLevel("WARN")
    //2.加载IP地址规则文件
    val ipFile: RDD[String] = sc.textFile("D:\\data\\ip.txt")
    //3.获取IP地址起始范围和结束范围、城市信息、经度、纬度
    val lineArr: RDD[Array[String]] = ipFile.map(_.split("\\|"))
    //RDD[(IP地址起始值,IP地址结束值,城市信息,经度,纬度)]
    val ipRuleRDD: RDD[(String, String, String, String, String)] = lineArr.map(x=>(x(2),x(3),x(4)+""+x(5)+""+x(6)+""+x(7)+""+x(8),x(13),x(14)))
    val ipRules: Array[(String, String, String, String, String)] = ipRuleRDD.collect()
    //注意:使用二分查找法时会去有序数组中查找指定元素的索引
    //注意:ipRules后续会被各个任务使用多次,为了避免多次传输,可以把它广播到各个Executor
    val ipRulesBroadcast: Broadcast[Array[(String, String, String, String, String)]] = sc.broadcast(ipRules)

    //4.加载日志文件
    val logFile: RDD[String] = sc.textFile("D:\\data\\20190121000132.394251.http.format")
    //5.将日志中的IP地址分割出来
    val ipRDD: RDD[String] = logFile.map(_.split("\\|")).map(_(1))
    /*6.将IP地址转为数字,并使用二分查找法去ipRules中查找出相应的城市信息,记为((城市,经度,纬度),1)*/
    val cityInfoAndOne: RDD[((String, String, String), Int)] = ipRDD.map(ip => {
      val ipRulesArr: Array[(String, String, String, String, String)] = ipRulesBroadcast.value
      //将IP地址转为数字
      val ipNum: Long = IPUtils.ipToLong(ip)
      //去ipRules中找出索引
      val index: Int = IPUtils.binarySerarch(ipNum, ipRulesArr)
      //根据索引去ipRules中获取城市、经度、纬度
      val t = ipRulesArr(index)
      //将(城市,经度,纬度)记为1,即返回((城市,经度,纬度),1)
      ((t._3, t._4, t._5), 1)
    })
```

```scala
//7.将((城市,经度,纬度),1)进行聚合得出统计结果
//RDD[((城市,经度,纬度), count)]
val cityInfoAndCount: RDD[((String, String, String), Int)] = cityInfoAndOne.reduceByKey(_+_)
cityInfoAndCount.collect().foreach(println)
//8.将结果数据写入MySQL中
val result: RDD[(String, String, String, Int)] = cityInfoAndCount.map(t=>
(t._1._1,t._1._2,t._1._3,t._2))
//result.foreachPartition(iter => IPUtils.save(iter))
//注意：函数式编程的思想是行为参数化
result.foreachPartition(IPUtils.save)

//println(IPUtils.ipToLong("192.168.100.100"))
}
object IPUtils{
    def binarySerarch(ipNum: Long, ipRulesArr: Array[(String, String, String, String, String)]): Int = {
        //二分查找
        var start = 0
        var end: Int = ipRulesArr.length -1
        while(start <= end){
          var middle: Int =  (start + end) /2
         val t = ipRulesArr(middle)
         val startIp: Long = t._1.toLong
         val stopIp: Long = t._2.toLong
         //如果传入的ipnum正好在开始和结束范围之内，则返回该索引
         if (ipnum >= startIp && ipnum <= stopIp){
           return middle
         }else if (ipnum < startIp){
           end = middle -1
         }else if (ipnum > stopIp){
           start = middle + 1
         }
        }
        -1 //如果循环结束没有找到则返回-1
    }

    //ipToLong方法接收一个String类型的IP地址，如192.168.100.100，返回一个数字，如3232261220
    def ipToLong(ip:String):Long ={
        //注意：IP地址原始面貌如下
```

```scala
        //10111111.10111010.11110000.11111100
        val ipArr: Array[Int] = ip.split("[.]").map(s => Integer.parseInt(s))
        var ipnum = 0L
        //原本ipnum为00000000.00000000.00000000.00000000
        //第一次循环结束ipnum为00000000.00000000.00000000.10111111
        //第二次循环ipnum左移,变为00000000.00000000.10111111.00000000
        //第二次循环结束ipnum为00000000.00000000.10111111.10111010
        //最后ipnum为10111111.10111010.11110000.11111100
        for(i <- ipArr){
          ipnum = i | (ipnum << 8)
        }
        ipnum
      }
      def save(iter: Iterator[(String, String, String, Int)]): Unit = {
        val conn: Connection = DriverManager.getConnection("jdbc:mysql://127.0.0.1:3306/bigdata", "root", "root")
        val sql: String = "INSERT INTO 'iplocaltion' ('id', 'city', 'longitude', 'latitude', 'total_count') VALUES (NULL, ?, ?, ?, ?);"
        val preparedStatement: PreparedStatement = conn.prepareStatement(sql)
        try {
          for (i <- iter) {
            preparedStatement.setString(1, i._1)
            preparedStatement.setString(2, i._2)
            preparedStatement.setString(3, i._3)
            preparedStatement.setLong(4, i._4)
            preparedStatement.executeUpdate()
          }
        } catch {
          case e:Exception => println(e)
        } finally {
          if (conn != null){
            conn.close()
          }
          if(preparedStatement != null){
            preparedStatement.close()
          }
        }
      }
    }
  }
}
```

2. Spark SQL 实现方式

① 创建 SparkSession。

② 加载 IP 地址规则文件。

③ 获取 IP 地址起始范围和结束范围、城市信息、经度、纬度。

④ 将 RDD 转换成 DataFrame。

⑤ 加载日志文件。

⑥ 获取日志中的 IP 地址并转为数字，最后将 RDD 转换成 DataFrame。

⑦ 注册表 t_iprules、t_ips。

⑧ 关联查询。

代码实现如下。

```scala
package cn.itcast.ip
import java.sql.{Connection, DriverManager, PreparedStatement}
import org.apache.spark.SparkContext
import org.apache.spark.rdd.RDD
import org.apache.spark.sql.{DataFrame, SparkSession}

object IPMap_SQL {
  def main(args: Array[String]): Unit = {
    //1.创建SparkSession
    val spark: SparkSession = SparkSession.builder().master("local[*]").appName("Spark SQL").getOrCreate()
    val sc: SparkContext = spark.sparkContext
    sc.setLogLevel("WARN")
    //2.加载IP地址规则文件
    val ipFile: RDD[String] = sc.textFile("D:\\data\\ip.txt")
    //3.获取IP地址起始范围(2)和结束范围(3)、城市信息(4,5,6,7,8)、经度(13)、纬度(14)
    val lineArr: RDD[Array[String]] = ipFile.map(_.split("\\|"))
    //RDD[(IP地址起始值，IP地址结束值，城市信息，经度，纬度)]
    val ipRuleRDD: RDD[(String, String, String, String, String)] = lineArr.map(x=>(x(2),x(3),x(4)+""+x(5)+""+x(6)+""+x(7)+""+x(8),x(13),x(14)))
    //4.将RDD转换成DataFrame
    import spark.implicits._
    val ipRulesDF: DataFrame = ipRuleRDD.toDF("startnum","endnum","city","longitude","latitude")
```

```
//ipRulesDF.show(10)
/*
+--------+--------+----------+---------+---------+
|startnum|  endnum|      city|longitude| latitude|
+--------+--------+----------+---------+---------+
|16777472|16778239|亚洲中国福建福州|119.306239|26.075302|
|16779264|16781311|亚洲中国广东广州|113.280637|23.125178|
|16785408|16793599|亚洲中国广东广州|113.280637|23.125178|
|16842752|16843007|亚洲中国福建福州|119.306239|26.075302|
|16843264|16844799|亚洲中国福建福州|119.306239|26.075302|
|16844800|16859135|亚洲中国广东广州|113.280637|23.125178|
|16908288|16908799|亚洲中国福建福州|119.306239|26.075302|
|16908800|16909055|亚洲中国北京北京海淀| 116.29812| 39.95931|
|16909312|16909567|亚洲中国北京北京|116.405285|39.904989|
|16909568|16910335|亚洲中国福建福州|119.306239|26.075302|
+--------+--------+----------+---------+---------+
*/
//5.加载日志文件
val logFile: RDD[String] = sc.textFile("D:\\data\\20190121000132.394251.http.format")
//6.获取日志中的IP地址并转为数字,最后将RDD转换成DataFrame
val ipDF: DataFrame = logFile.map(_.split("\\|")).map(arr=>IPUtils.ipToLong(arr(1))).toDF("ipnum")
//ipDF.show(10)
/*
+----------+
|     ipnum|
+----------+
|2111136891|
|1969608581|
|1969610308|
|1937253494|
|2076524791|
|3728161200|
|2076525149|
|1937247389|
|1937246192|
|1969609713|
+----------+
*/
//7.注册表t_iprules、t_ips
```

```scala
    ipRulesDF.createOrReplaceTempView("t_iprules")
    ipDF.createOrReplaceTempView("t_ips")
    //8.关联查询
    val sql:String =
      """
        |select city,longitude,latitude,count(*) counts
        |from t_iprules
        |join t_ips
        |on ipnum >= startnum and ipnum <= endnum
        |group by city,longitude,latitude
        |order by counts desc
      """.stripMargin
    spark.sql(sql).show
    /*
    +------------------+----------+---------+------+
    |              city| longitude| latitude|counts|
    +------------------+----------+---------+------+
    |    亚洲中国陕西西安|108.948024|34.263161|  1824|
    |    亚洲中国北京北京|116.405285|39.904989|  1535|
    |    亚洲中国重庆重庆|106.504962|29.533155|   400|
    |  亚洲中国河北石家庄|114.502461|38.045474|   383|
    |  亚洲中国重庆重庆江北| 106.57434| 29.60658|   177|
    |    亚洲中国云南昆明|102.712251|25.040609|   126|
    |亚洲中国重庆重庆九龙坡| 106.51107| 29.50197|    91|
    |  亚洲中国重庆重庆武隆|  107.7601| 29.32548|    85|
    |  亚洲中国重庆重庆涪陵| 107.39007| 29.70292|    47|
    |  亚洲中国重庆重庆合川| 106.27633| 29.97227|    36|
    |  亚洲中国重庆重庆长寿| 107.08166| 29.85359|    29|
    |  亚洲中国重庆重庆南岸| 106.56347| 29.52311|     3|
    +------------------+----------+---------+------+
    */
}
object IPUtils{
    //ipToLong方法接收一个String类型的IP地址,如192.168.100.100,返回一个数字,如3232261220
    def ipToLong(ip:String):Long ={
      //注意:IP地址原始面貌如下
      //10111111.10111010.11110000.11111100
      val ipArr: Array[Int] = ip.split("[.]").map(s => Integer.parseInt(s))
      var ipnum = 0L
      //原本ipnum为00000000.00000000.00000000.00000000
```

```
          //第一次循环结束ipnum为00000000.00000000.00000000.10111111
          //第二次循环ipnum左移，变为00000000.00000000.10111111.00000000
          //第二次循环结束ipnum为00000000.00000000.10111111.10111010
          //最后为10111111.10111010.11110000.11111100
          for(i <- ipArr){
             ipnum = i | (ipnum << 8)
          }
          ipnum
      }
   }
```

也可以通过广播变量和 UDF 的方式完成该项目，代码如下。

```
package cn.itcast.ip
import org.apache.spark.SparkContext
import org.apache.spark.broadcast.Broadcast
import org.apache.spark.rdd.RDD
import org.apache.spark.sql.{DataFrame, SparkSession}

object IPMap_SQL_One {
   def main(args: Array[String]): Unit = {
      //1.创建SparkSession
      val spark: SparkSession = SparkSession.builder().master("local[*]").appName("Spark SQL").getOrCreate()
      val sc: SparkContext = spark.sparkContext
      sc.setLogLevel("WARN")
      //2.加载IP地址规则文件
      val ipFile: RDD[String] = sc.textFile("D:\\data\\ip.txt")
      //3.获取IP地址起始范围和结束范围、城市信息、经度、纬度
      val lineArr: RDD[Array[String]] = ipFile.map(_.split("\\|"))
      //RDD[(IP地址起始值，IP地址结束值，城市信息，经度，纬度)]
      val ipRuleRDD: RDD[(String, String, String, String, String)] = lineArr.map(x=>(x(2),x(3),x(4)+""+x(5)+""+x(6)+""+x(7)+""+x(8),x(13),x(14)))
      val ipRulesArr: Array[(String, String, String, String, String)] = ipRuleRDD.collect()
      //把IP地址规则作为广播变量发送到各个Executor，方便一个Executor中的多个任务共享
      val ipBroadcast: Broadcast[Array[(String, String, String, String, String)]] = sc.broadcast(ipRulesArr)
      import spark.implicits._
      //4.加载日志文件
      val logFile: RDD[String] = sc.textFile("D:\\data\\20190121000132.394251.http.format")
```

```
//5.获取日志中的IP地址并转为数字,最后将RDD转换成DataFrame
val ipDF: DataFrame = logFile.map(_.split("\\|")).map(arr=>IPUtils.ipToLong
(arr(1))).toDF("ipnum")
//ipDF.show(10)
/*
+----------+
|     ipnum|
+----------+
|2111136891|
|1969608581|
|1969610308|
|1937253494|
|2076524791|
|3728161200|
|2076525149|
|1937247389|
|1937246192|
|1969609713|
+----------+
*/
//6.注册表t_iprules、t_ips
ipDF.createOrReplaceTempView("t_ips")
//7.UDF,传入ipnum返回城市信息
spark.udf.register("ipnumToCityinfo",(ipnum:Long)=>{
    val ipRules: Array[(String, String, String, String, String)] = ipBroadcast.value
    val index: Int = IPUtils.binarySerarch(ipnum,ipRules)
    val t: (String, String, String, String, String) = ipRules(index)
    t._3 + t._4 + t._5
})
//8.关联查询
val sql:String =
    """
      |select ipnumToCityinfo(ipnum) cityinfo,count(*) counts
      |from t_ips
      |group by cityinfo
      |order by counts desc
    """.stripMargin
spark.sql(sql).show(truncate = false)
/*
+-----------------------------+------+
|cityinfo                     |counts|
```

```
    +------------------------------+------+
    |亚洲中国陕西西安108.94802434.263161   |1824 |
    |亚洲中国北京北京116.40528539.904989   |1535 |
    |亚洲中国重庆重庆106.50496229.533155   |400  |
    |亚洲中国河北石家庄114.50246138.045474 |383  |
    |亚洲中国重庆重庆江北106.5743429.60658 |177  |
    |亚洲中国云南昆明102.71225125.040609   |126  |
    |亚洲中国重庆重庆九龙坡106.5110729.50197|91  |
    |亚洲中国重庆重庆武隆107.760129.32548  |85   |
    |亚洲中国重庆重庆涪陵107.3900729.70292 |47   |
    |亚洲中国重庆重庆合川106.2763329.97227 |36   |
    |亚洲中国重庆重庆长寿107.0816629.85359 |29   |
    |亚洲中国重庆重庆南岸106.5634729.52311 |3    |
    +------------------------------+------+
     */

  }
  object IPUtils{
    def binarySerarch(ipNum: Long, ipRulesArr: Array[(String, String, String,
String, String)]): Int = {
      //二分查找
      var start = 0
      var end: Int = ipRulesArr.length -1
      while(start <= end){
        var middle: Int =  (start + end) /2
       val t = ipRulesArr(middle)
       val startIp: Long = t._1.toLong
       val stopIp: Long = t._2.toLong
        //如果传入的ipnum正好在开始和结束范围之内，则返回该索引
        if (ipnum >= startIp && ipnum <= stopIp){
          return middle
        }else if (ipnum < startIp){
          end = middle -1
        }else if (ipNum > stopIp){
          start = middle + 1
        }
      }
      -1 //如果循环结束没有找到则返回-1
    }

    //方法接收一个String类型的IP地址，如192.168.100.100，返回一个数字，如3232261220
```

```scala
def ipToLong(ip:String):Long ={
  //注意：IP地址原始面貌如下
  //10111111.10111010.11110000.11111100
  val ipArr: Array[Int] = ip.split("[.]").map(s => Integer.parseInt(s))
  var ipnum = 0L
  //原本ipnum为00000000.00000000.00000000.00000000
  //第一次循环结束ipnum为00000000.00000000.00000000.10111111
  //第二次循环ipnum左移，变为00000000.00000000.10111111.00000000
  //第二次循环结束ipnum为00000000.00000000.10111111.10111010
  //最后ipnum为10111111.10111010.11110000.11111100
  for(i <- ipArr){
    ipnum = i | (ipnum << 8)
  }
  ipnum
}
```

11.3 本章总结

本章主要介绍了网站运营指标统计项目和 IP 经纬度热力图分析项目，从需求分析、数据分析到最后代码开发，都进行了详细的讲解。另外，在项目开发过程中，我们充分体验了 Spark 的 RDD、Spark SQL 等技术的应用。

11.4 本章习题

1. 在本地完成网站运营指标统计项目。

2. 在本地完成 IP 经纬度热力图分析项目。